unlocking specialist potential

a more
ırticipative
role for
specialist
:ontractors

by
of England
and
of England

r Task Force
:tion Forum

nvironment

RCF

READING CONSTRUCTION FORUM

unlocking: specialist potential

a more participative role for specialist contractors

Published by: Reading Construction Forum
PO Box 219
Whiteknights
Reading
Berkshire RG6 6AW
Tel: 0118 931 8766
Fax: 0118 931 3856

Text Editor: Alan Spence

Designed by: Rufus Stone Design and Public Relations Limited
Tel: 01635 551404

Reprographics: HBS Design Associates Limited
Tel: 01635 33301/528145

Printed by: Chromocraft Limited
Tel: 0181 7514341

ISBN: 1 902266 00 5

British Library Cataloguing in Publication Data. A catalogue record for this publication is available from the British Library

contents

Foreword by Sir Michael Latham
Author, "Constructing the Team" and Chairman of
The Construction Industry Board, 1995-96

Foreword

I warmly welcome this important study of how to unlock specialist contractor potential. In a rapidly changing industry environment, both clients and the supply side are increasingly looking to improve performance and reduce, and then hopefully eliminate, conflict and disputes through a teamwork approach. "Constructing the Team" and all the detailed work which has followed from it over the last four years have had three key concepts as their driving forces. The first was that the client was at the core of the process. The second was that the route to achieve that was teamworking to produce a step change in performance improvement. the third was that partnering was a very effective method to reach that goal. Overseas experience, especially in the United States, has been very encouraging. We are now beginning to build up our own case histories here in the United Kingdom.

The specialist contractor has a crucial role to play in delivering client satisfaction. His/her work adds vital value to the project, but best performance is achieved through early involvement of the specialist contractor's expertise, particularly as it involves a significant amount of detailed design. This study rightly insists that all sectors of the construction process have a role to play in improving performance. Everyone needs to improve their own internal procedures and train their staff to accept enthusiastically a new method of working. That is not easy or painless. The report sets out a clear action plan towards the end, with an executive summary at the beginning. You should, of course, carefully read the whole report. But the action plan should be pinned up in each office, indeed on each site, to focus attention on the continual need for teamworking.

I warmly commend the report and congratulate its authors and all those involved in its conception and production.

Michael Latham

Michael Latham
April 1998

Construction, like many other sectors of the UK economy, is changing to meet demands for greater competitiveness.

Executive Summary

If the construction industry is to successfully embrace the changes needed to improve performance, the specialists need to play a more significant role, given that they account for as much as 80% of contract expenditure.

Other sectors of the economy such as aerospace, retailing and automotive industries are already effectively implementing a more participative role for their specialist contractors and suppliers.

To enable the specialists within the design and construction process to play a key role, this report proposes three strategies to unlock their potential for adding value:

- improving relationships through teamwork and collaboration.
- taking a process orientated approach to the design and construction process.
- making customer requirements the central focus.

These strategies are as applicable to "design and build" projects as they are to those completed on a more traditional basis. They are both radical and logical, and are already helping businesses in a number of sectors of the economy to reach world-class capability and competitiveness.

Organisations that have adopted only elements of this approach have already identified bottom-line rewards such as labour savings of 20%, a 70% reduction in stockholding, a 20% reduction in waste, zero defects, reduced cycle time, as well as less tangible improvements in flexibility and teamwork.

The strategies proposed in this report could result in a 30% reduction in costs as well as substantial enhancement in the overall performance of the construction industry. However, to be fully effective, they need the full support of all those involved throughout the whole process.

There will have to be a cultural change which can only be achieved and sustained through continuous and shared learning in relation to aspects of the whole process. For these changes to be permanent the learning must permeate all levels of all organisations associated with the design and construction of projects.

The diffusion of change throughout the whole industry will be achieved through the transfer of the learning and the sharing of good practice between projects and organisations.

Experience from other sectors indicates that successful implementation of such strategies is dependent on strong leadership with one of the partners acting as a champion of change.

Methodology

The objective of the project was to investigate the ways in which the performance of specialist contractors in construction could be improved through their own actions and those of other organisations involved in the construction process.

The methodology combined qualitative and quantitative approaches and was influenced by the need to gain an insight into the functioning of the overall process of design and construction and the barriers to improving performance.

The project comprised four major stages:

- in-depth interviews were undertaken with thirty specialist contractors in order to elicit the problems they were experiencing and the implications for the process, and to identify emerging approaches to improve their performance
- the findings from the in-depth interviews were checked against a larger sample of specialist contractors and main contractors and design consultants.
- the findings from that fieldwork were compared with the approaches being used by other sectors (automotive, retailing and aerospace) to address similar problems.
- a series of workshops were held with groups of practitioners in order to ensure the relevance, practicality and implementation potential of the research findings.

The Research Model

The methodology also included the development of a model which reflected the complexities of the relationships and processes associated with construction. The simple strategic model was used to benchmark effectively across projects and organisations in construction and other sectors.

Objectives of the model:

- represent the key stages and processes of construction projects relevant to the investigation.
- identify the critical elements contributing to performance including strategy and policy, organisational structures, management, technology, resources, leadership, processes and people.
- highlight the essential role of people, their inter-relationships and task responsibility
- incorporate the critical project success factors including customer satisfaction (internal and external), and business results.

This model was seen as being necessary to represent current practices and to highlight the major problems being encountered by specialist contractors. In a later stage of the research project the model was used to help identify the emerging strategies that are being used to improve performance.

The chosen model was derived from a number of models used to represent processes and assess performance within construction and other sectors of the economy. The European Quality Award Model was considered particularly relevant to this project for the following reasons:

- focus on customer satisfaction.
- simplicity yet comprehensiveness.
- wide acceptability across sectors.
- easily adaptable to the needs of the research project.
- focus on inputs, processes and outputs as well as enablers and results.
- suitability to assess corporate as well as project performance.
- relevance to the role of critical enabling elements (people, leadership, management, policy and strategy, resources) and results (customer satisfaction and business results).

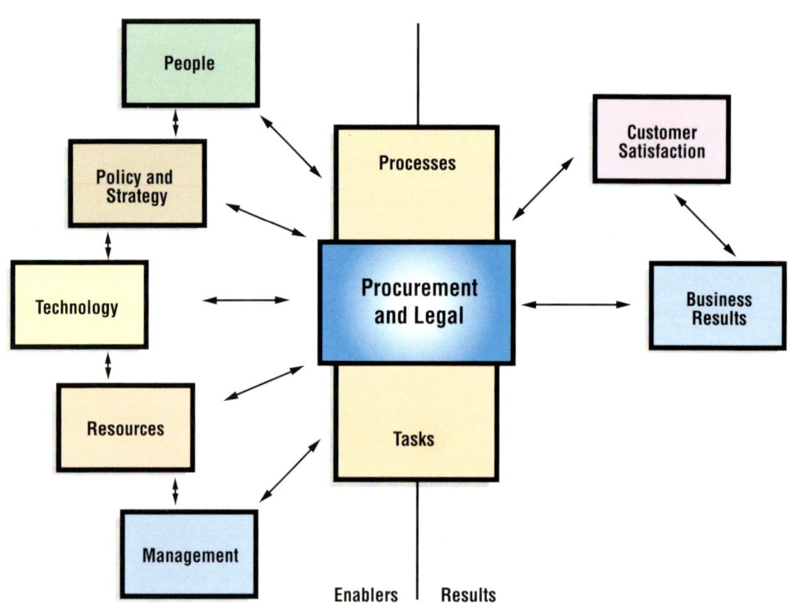

The model for this research project

Acknowledgements

The authors would like to thank the following people for their help in compiling this report: all the Task Force members:

Peter Appleby	-	Spen Hill Properties/Tesco
Richard Saxon	-	BDP
Tony Speroni	-	Coverite
Michael Wilson	-	Birse
Ray O' Rourke	-	O' Rourke Group
Graeme Telford	-	SmithKline Beecham
Peter Frackiewicz	-	Land Securities
David Bucknall	-	Bucknall Group

and the Chair of the Task Force, Martin Davis of Drake & Scull for their clear direction and commitment to the report. All have consistently helped our work with advice and practical support.

We would also like to acknowledge the valuable insights provided by:

David Blackbrough, Automotive Directorate, DTI, Birmingham.

Sue Coss, Group Human Resources Development Manager, Bristol and West Building Society.

John Cavanagh, Continuous Improvement Manager, Procurement Department, British Aerospace.

Philip Knight, Group Purchasing, Rover Group.

David Millar, Buying Controller, Tesco.

Sue Parker, Automotive Directorate, DTI, Birmingham.

Sarah Pennington, Supply Chain Executive, Somerfield.

Mike Wilkins, Deputy Director, Partnership Sourcing Ltd.

Centre for Purchasing and Supply Management, Bath School of Management, University of Bath.

Centre for Research in Innovation and Management, Brighton University.

The Lean Enterprise Centre, Cardiff Business School, University of Wales.

The Logistics Systems Dynamics Group, University of Wales, Cardiff.

The Research Unit for Defence Economics, The Faculty of Economics and Social Science, University of the West of England, Bristol.

Professor J.S. Morrison and naval architect Mr. J.S. Coates for picture reference on the Greek Trireme.

The authors greatly acknowledge the valuable contribution provided by Alan Rogan, a research student sponsored by the Steel Construction Institute, in the initial data collection phase of the project.

The authors are founding members of INTERFORM@UWE, a multi-disciplinary research group based at the University of the West of England in Bristol. INTERFORM seeks to identify and transfer best practice in management and technology across a number of sectors of the economy.

Reading Construction Forum

The Reading Construction Forum is based at the University of Reading. It has close links with the Department of Construction Management & Engineering but is independent of the University and regularly commissions research from other universities such as the University of the West of England at Bristol.

Members of the Forum are major companies concerned with achieving high quality in the design, construction and use of commercial, retail and industrial buildings. All are committed to change and innovation in the British and European building industries.

To achieve effective action the Forum is led by Principal Members who are drawn from international customers and building industry consultants, contractors and manufacturers. They work with leading researchers to identify specific actions needed to improve the performance of the UK building industry. However, to ensure that new ideas and best practice are applied widely, the Forum includes, as Members, organisations with varying levels of experience that want to take practical actions aimed at achieving high quality in building.

The over-riding purpose of the Forum is to enlist the active involvement of members in identifying, debating and activating the objectives.

The objectives are designed to achieve change by collaborative involvement rather than by perpetuating the confrontational aspects of the industry that historically have damaged its performance. To these ends the objectives of the Forum are:

- to provide a forum for members about all issues affecting quality, efficiency and innovation in the design, construction and use of commercial and industrial buildings.
- to identify the need for specific changes.
- to ensure that good practice which unites the industry in the interests of customers is widely disseminated.
- to encourage the research and development needed to ensure that the changes are effective and beneficial.
- to take practical actions aimed at improving the international competitiveness of the building industries of Britain and Europe, including helping to improve relationships between Industry and Government.
- to implement the results of research by the Reading Construction Forum
- to work closely with other organisations committed to changing the construction industry for the better.

The Specialist Contractor Task Force was established in 1996 to assist the research team from the University of the West of England in its research into the problems faced by specialist contractors and into the means to overcome them.

The Specialist Contractor Effectiveness Task Force consists of:

Martin Davis (Chairman)	Drake & Scull
Peter Appleby	Spen Hill Properties/Tesco
Richard Saxon	BDP
Tony Speroni	Coverite
Michael Wilson	Birse
Ray O'Rourke	O'Rourke
Graeme Telford	SmithKline Beecham
Peter Frackiewicz	Land Securities
David Bucknall	Bucknall Group

The additional assistance of Bernard Dempsey (O'Rourke), Kit Moore (SmithKline Beecham), Peter Snape (BDP) and David Wheeler (Birse) on the Task Force has been greatly welcomed.

The Reading Construction Forum firmly believes in the need to increase the effectiveness of specialist contractors within the design and construction process and will be developing further the recommendations within this report.

The Forum consists of the following companies:

Principal Members

Bovis Europe
Building Design Partnership
Roger Bullivant Ltd
Drake & Scull
Gardiner & Theobald
John Laing plc
John Lewis Partnership
O'Rourke Group
Richard Rogers Partnership
ROM Limited
J. Sainsbury plc
Slough Estates plc
Stanhope plc
Tarmac
Waterman Partnership

Members

Andover Controls
Barclays Property Holdings
B & Q plc
Birse Construction
Bucknall Group
Coverite Ltd
Ernest Green & Partners
Galliford (UK) Ltd
Gleeds
M J Gleeson Group plc
Hochtief
Irvine Whitlock Ltd
Kyle Stewart
Land Securities plc
Lovell Construction
Mansell
John Mowlem Construction plc
Nabarro Nathanson
Otis plc
R G Construction Management
Shepherd Construction
SmithKline Beecham
Stent Foundations
Taylor Joynson Garrett
Tesco Stores
Try Construction
Union Railways
Wates Integra

The need for change

Becoming World Class

There is a widespread and growing awareness that the UK construction industry needs to be more efficient and competitive.

Becoming World Class

The Latham Review[1] and the Technology Foresight Programme for Construction[2] are among many influential calls for action to reduce waste, give customers better value and produce world-class products and services.

As the Technology Foresight Programme points out, the UK enjoys some of the lowest construction input costs in Europe, in terms of labour and materials, but some of the highest unit output prices in terms of buildings and structures. Customers, in particular, are demanding better value.

There have been many positive responses from within the industry to the need for improvement. But much remains to be done.

"The competitive advantages which we thought we had are declining. The competitiveness of the construction industry needs improvement."

Herb Nahapiet, Construction Panel Chairman, Technology Foresight.

A key role for specialist contractors

This report aims to contribute to the process of change in order to improve performance by focussing on the role of specialist contractors.

The shift from direct employment to sub-contracting and specialist contracting which characterises all sectors of the economy[3] is particularly significant in shaping business processes in the construction industry. It has resulted in a more substantial role for the specialists. Given that they now account for as much as 80 percent of contract expenditure, the specialists clearly can and should play a leading part in helping the industry to adopt the new approaches needed to improve the overall performance of the design and construction process. The Latham Report points to their largely untapped potential to contribute to the target of reducing construction costs by 30 percent by the year 2000.

What is world class?

The Government White Paper, 'Competitiveness - Creating the Enterprise Centre of Europe', refers to the 'Winning Report'[4], a CBI/DTI study of 120 of the UK's most successful companies selected because of their profitability, growth, and recognised standards in their sector.

It sets out the characteristics which distinguish the leading edge or world class from the run-of-the mill.

- they have strong leaders who champion change, set targets and are open with customers, staff, investors and suppliers.
- the potential of their people is unlocked by simplifying management, developing skills and encouraging working in teams.
- they constantly learn from others and are committed to innovation and the continuous search to introduce new products and services by exploiting new technologies and other ways of securing a competitive edge.
- they anticipate customers' requirements and consistently exceed their expectations.

Successful firms also exploit the inter-relationships between these characteristics. For example, by empowering and training employees they encourage innovation; by fostering good design and marketing, they focus on the needs of the customer.

A study of 560 manufacturing companies by the TEC National Council[5], also quoted in the Government White Paper, and a survey of 115 UK service sector companies[6] reached similar conclusions.

Less successful companies invariably lack one or more of the basic characteristics, and this highlights the importance of an holistic approach which balances the needs of all involved in building business success.

Unlike other industries, construction has so far done little to give specialists a more central role in its efforts to improve effectiveness and become more competitive.

Figure 1 A more participative role for specialist contractors

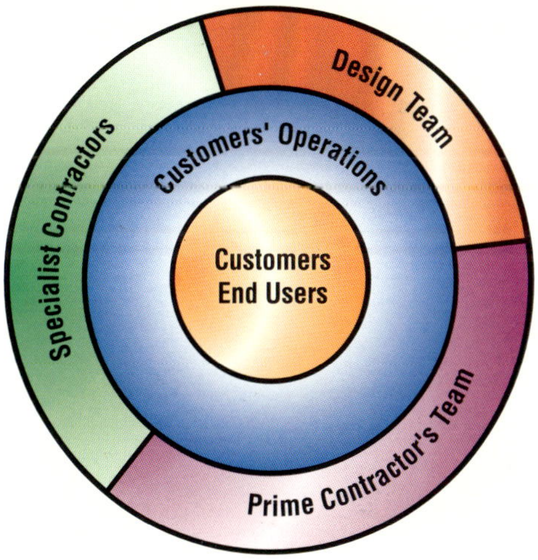

Factors inhibiting specialist contractors

Interviews with participants in the design and construction process underlined the strong 'them and us' environment of mistrust and hostility in which consultants and contractors criticise the poor performance of the 'subbies', while the specialists feel that they are being undervalued by the consultants and contractors.

The adversarial culture and traditions of the industry lead to disputes, conflict and a lack of trust between project partners. Among the most significant features of this malaise are:

- unrealistic project programmes.
- undue emphasis on cost rather than value, producing fierce competition and reduced margins.
- dutch auctions.
- perceived poor status of specialist contractors.
- onerous contract conditions and inappropriate unloading of risk.
- unreasonable payment procedures.
- lack of understanding of the risks involved and their consequences.
- unclear statements of requirements and ambiguous project information and tender packages.
- insufficient focus on customer–supplier relationships.

The relationship and links between Intel and computer manufacturers

As soon as Intel was ready to market the new 286, 386 and 486 series of microprocessor, computer manufacturers had designs based on the new chip ready to go into production.

The computer companies do not wait for Intel to complete development and then release design information. Nor do they design systems and then write a microprocessor specification for Intel.

The manufacturers work closely with their supplier, Intel, and highly reliable products emerge after short development cycles.

Source: Partnership Sourcing

Leyland Trucks aims to become world class

Leyland trucks has been through a long process of reorganisation in response to a changing marketplace and changes in ownership. It has emerged with a highly respected culture that has been developed throughout the whole workforce. This has required the participation of all employees. But it has been led by a dedicated, committed and forward-looking management team.

Four cornerstones have been set for running the business:

training

team building

involvement

empowerment

The most significant asset on the road to world-class competitiveness has been the dedication and imagination of the management team. This has encouraged a strong sense of teamwork throughout the company, resulting in many situations that once caused friction and misunderstanding between management and the shop floor being resolved by the workers themselves - to the benefit of all sides.

Leyland Trucks is keen to share its experiences and has set up a learning centre that considers the theory and then looks at its practical application, through the experiences of practitioners

Source: Partnership Sourcing.

- lack of clarity of agreed requirements with customers and suppliers.
- irregular flow of work.
- insufficient project planning resulting in overlapping of functions and waste.
- poor flow of information.
- underdeveloped management and interpersonal skills.

Squeezing the margins

In construction, competition is based almost solely on price. Time plays a role, as does quality, but contracts are most frequently awarded to the lowest bidder.

Traditionally, specialists supplying systems such as mechanical and electrical services find themselves squeezed between prescriptive specifications under which the designer inhibits flexibility and innovation on the one hand and, on the other, manufacturers that have been specified by the designer and can therefore command their price. This leaves inadequate margins for the engineering and management input the specialist is expected to supply. This has in turn provided a motive to economise and cut corners wherever possible, with the inevitable conflict with the customer's objectives.

A better way forward

Over the past decade more and more companies have recognised the weaknesses of the industry's adversarial, cost-driven approach. There is an increasing awareness that it must be more beneficial for all the participants in the design and construction process to work together in cooperation, recognising that each of them needs to make a fair profit whilst providing good value. Many companies have therefore sought to develop closer and more harmonious links with their specialist contractors and suppliers.

This approach is much more strongly developed in the automotive, aerospace and retailing industries. They have been successfully adopting new ways of working to help them achieve more productive and efficient relationships within their supply chains.

This report draws on experience of the new approaches within the construction industry and in other industries to propose a more participative role for the specialists, within a new framework.

"The industry must recognise that specialist contractors are major players in a construction project and involve them at a very early stage. An attitude of trust built on a teamwork approach must prevail on individual projects and subcontractors must play their part in similar fashion. They must also work to put their own houses in order to ensure that the cultural change is filtered down the line to the operatives on site to ensure they have a good working spirit and pride in the work they are carrying out at site level."

Peter Appleby, Construction Manager, Tesco Stores

Unlocking the potential

Unlocking the potential of the specialists can lead to greater mutual competitiveness and improvements in the whole design and construction process.

As in other sectors, their earlier involvement enables them to contribute their specialist expertise and experience towards meeting the rapidly changing needs of their customers.

This will not only enhance quality of service and coordination between different partners but will also reduce transaction costs, waste, overlapping and cycle time.

"Clearly by achieving earlier involvement there is the opportunity for developing the significant element of trust as well as providing and promoting the mechanics for partnering."

Paul Dewick, Director, Irvine Whitlock Ltd

This more co-ordinated and integrated approach can only be achieved if it is embraced by all organisations involved in the network of customers and suppliers within the design and construction process.

Many specialists are small firms and consequently have many strengths but also vulnerabilities which need to be recognised by the other participants. Experience from the other sectors indicates that the other participants must play their part in developing the strengths of the specialists whilst at the same time helping them to address their vulnerabilities.

The complexity of the design and construction process and the diversity, discreteness and uniqueness of its projects sets it apart in some respects from the processes associated with other industries. It embraces contracting firms, specialist contractors, consulting architects and engineers, quantity surveyors, suppliers of building materials and manufacturers of equipment as well as customers. It also involves a wide range of activities including the planning, regulation, design, manufacture, construction, maintenance and eventual decommissioning of buildings and other structures.

Nissan uses partnerships to create world-class UK suppliers

Japanese car manufacturer, Nissan, was one of the first companies in the automotive industry to introduce close relationships with its UK suppliers.

Partnerships were seen as the only way to compete with more established UK vehicle manufacturers and raise the performance of UK component suppliers to world class.

Since Nissan introduced Supplier Development Teams in 1988, over sixty suppliers have participated in activities, covering issues such as line efficiency, quality improvement, easy working, and autonomous maintenance.

Sommer has supplied Nissan with carpets and soft interior trim since 1984. At the heart of their partnership is a facility which Sommer opened near to Nissan which dramatically improved delivery times through synchronised Just In Time (JIT) delivery slots. There is no waiting stock and no inventory, cutting overheads for both partners.

Nissan maintain that such a close and dedicated partnership is possible because Sommer has a strong innovation capability. In 1993 the company won Nissan's most-improved supplier award. The following year they won the ultimate best supplier award. Sommer welcomes the change to its business that working with Nissan has brought.

Source: Partnership Sourcing

Unlocking Q Flooring's potential through earlier involvement

The benefits of earlier involvement of Q Flooring Systems through partnering arrangements with customers include:

- maximum cost savings.
- high quality installation for the customer.
- better planned workload for the specialist.

Rover works with its suppliers to add value while reducing costs

Rover uses supplier cost modelling to help keep supplier prices down without sacrificing quality. A cost model represents the costs suppliers incur to manufacture the product the customer buys.

Rover operates on the belief that a supplier is entitled to a fair profit. It helps both its buyers and its suppliers to see where the true costs are and what can be done to control them.

Because of the significant number of stages and participants, from both the public and private sectors, involved in the process, construction project outcomes are often considered difficult to predict. This may make the introduction of a more integrated approach more difficult. But it does not make it impossible.

Working more effectively to meet the challenge of a more integrated approach would assist the UK construction industry and the specialist contractors to acquire the features of world-class performance.

"A central feature of world-class organisations is the degree to which external processes and links with suppliers are developed. If the links are well developed and suppliers are made to be part of the team then these same suppliers will regard the particular world-class organisation as one they consider to be a good partner." Peter Hines[7]

Becoming world class

To improve the contribution of the specialists to the design and construction process and to raise their performance to world-class standards, this report proposes the adoption of the following three strategies:

- improve relationships through teamwork and collaboration
- take a process-orientated approach to design and construction
- make customer requirements the central focus of activity

Whilst these strategies are not new, if used to unlock the specialists' potential their significance will be transformed.

Design and build

These strategies are patently relevant to the traditional procurement methods under which the customer appoints the consultants for design and the contractor for construction, thereby artificially splitting what should be a continuous process. They are however equally relevant to the "design and build" route, under which an integrated approach between the consultants and the contractor is essential to the success of the overall project.

Strategies for Improvement

Teamwork and Collaboration

A Process Approach

Customer Focus

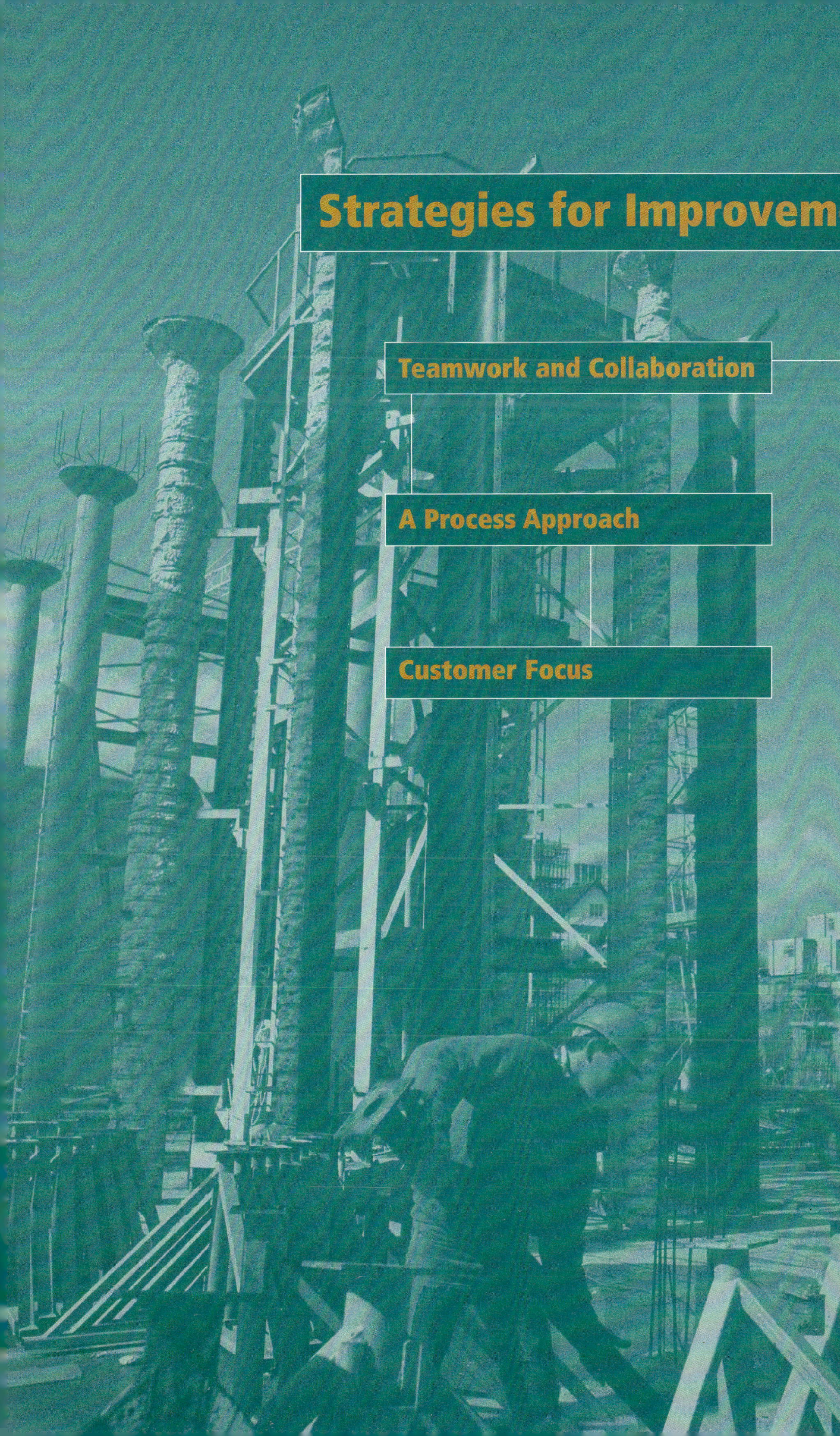

Teamworking is the key which can unlock the potential of specialists and improve their performance.

Teamwork and Collaboration

A Greek trireme under the traction of 170 oars in unison.

The unequal status accorded to the specialist contractors and suppliers and the unfair terms often imposed on them are part of the construction industry's culture of awarding contracts almost solely on the basis of price. The specialists are given little incentive to improve performance and add value.

It has been demonstrated within the construction industry, and to a greater extent in the automotive, aerospace and retailing industries, that there are better and more productive ways to organise relationships with suppliers and contractors

Good relationships through teamworking can be used, for instance, to find the best achievable way to realise the customers' objectives and the consultants' ideas. That said, there must of course be a framework which allows cost to be controlled. This can be achieved through contractors and specialists working together at all stages, starting with the design. Teamwork also allows open book accounting with costs open to scrutiny, profit levels declared and price related to the product or service provided.

In the case of M&E procurement, teamwork and collaboration can for instance reduce the disruption caused by the separation of the "scheme design" from the "detailed design."

Ideally the design process should be a continuum. The specialists need to be involved early and become part of the design team, initially acting as adviser to the consultant who prepares the design up to the scheme design stage. If the design process is really seamless, innovation, rationalisation and value engineering are genuinely possible.

However, in construction there is much less emphasis on teamworking than in other industries.

"In construction, manpower can account for up to 85% of total costs, against a typical figure of 25% in manufacturing, yet manufacturing devotes far more effort to team building."

Martin Chambers, Mace Ltd, Construction Productivity Network Workshop, Team Building to Improve Productivity on Complex Projects.

Specialist contractors are not viewed as equal partners in a project. They are criticised for lack of awareness of the need to raise their own capabilities. Bringing specialist contractors into the team will increase their awareness of the need to raise their own capabilities.

This requires a shift from the current "**sub**standard", "**sub**ordinate", and "**sub**missive" view of specialist contractors.

How specialist contractors view their relationships with other project partners

"The problems start with the relationship between the customer and the architect. The customer employs the architect in isolation to design what he thinks he wants. He chooses the architect by some form of competitive tender. So right from the start, instead of a partnership being set up, with all the members of the building team, we have a contract of commercial necessity. This inauspicious start is amplified at each stage of the tendering process until it gets to the people who actually do the work, the specialists, when sometimes the original expectations are no longer even recognisable. This all happens because the completion of the project within the budget is more important than the quality of the build."

"Most main contractors still perceive it as being in their own interest to continue with adversarial attitudes towards specialist contractors."

"It doesn't matter how clear the payment terms are, if they don't want to pay they won't."

"Problems on site will continue until there is trust between all parties and each party prices the contract at a price they can afford to carry out the work without relying on an income from exploiting the failure, or alleged failure, of others."

"As long as the main contractor and/or customer end of the industry continues to push for the absolute lowest price, and the specialist end of the industry accepts that position, the industry will be dogged by contractors who cut corners, employ poor standards and use labour that is either unqualified or so called unemployed."

"Partnering is where a group of contractors get into bed together - the trouble is the main contractor usually has all the bedclothes."

"Poor financial rewards have resulted in poor relationships between main contractors and specialist contractors and a negative culture has developed."

Turning groups into teams

In construction, individuals seldom work in isolation from others. Construction companies and projects are made up of groups of people and these groups exert a powerful influence over the way their members behave and perform their tasks. The challenge is to transform construction groups into **teams** which collaborate more effectively. This can be achieved by:

- the strengths of the specialist contractors being recognised and involved as early as possible in projects.
- the industry capitalising on the good informal relationships which often exist at the operational level on site.
- specialist contractors taking a more pro-active role by putting more emphasis on developing their communication with other project partners.
- sharing profits from the cost savings created by teamworking in order to reward appropriate involvement and provide incentives to the further development of teamworking.

No easy answers

Teamwork is no quick and easy fix. It will not work without a new culture based on trust. The industry needs to develop:

- better relationships between individuals and organisations based on trust and mutual understanding.
- 'open book' procedures and clearer payment terms.
- a more modern and non adversarial form of contract.
- a more appropriate format for dispute resolution.

In construction, individuals seldom work in isolation from others. The challenge is to transform construction groups into teams.

- improved coordination and exchange of information.
- greater emphasis on the development of interpersonal skills.
- feedback within companies and between companies.

Teambuilding requires effort and care

Teambuilding takes time and resources and must not be rushed. It is important to allow team members to define and allocate roles and develop trust and cohesiveness, to establish norms and practices before they can be fully effective and productive.[8] This, of course, presents difficulties within construction projects where time constraints often apply. However, the investment of time will provide tangible returns in terms of motivation, commitment, learning and performance.

To be fully effective, teams need to be fairly autonomous groups with broad terms of reference and limited supervision. They designate the positions to be filled and the allocation of members.

At the same time care must be taken to ensure that they are not given inappropriate tasks or impossible constraints, that the right number of the right type of people are appointed to them, and that they focus on solving problems, not just defining them.

"That teamworking is growing cannot be doubted. A survey by the Industrial Society of 500 personnel managers found that 40 per cent worked in organisations with self-managed teams. The average team in this survey had around eight people and the main reasons cited for their use were improved customer service, increased staff motivation, and quality of output."

Crainer[9]

Benefits of teamworking

Teamwork can deliver powerful benefits to the design and construction process:

- teams can be empowered and highly motivated.
- they are able to respond quickly to customer need, to develop cross functional links and to take advantage of opportunities.
- they can develop strong 'esprit de corps' and compensate for limited career paths in flatter organisations.
- they should form one of the primary mechanisms by which mutually agreed objectives are achieved.

Tesco recognise the value of teamworking

Tesco sees teamwork as one of the most important factors in the construction process. It maintains that the construction industry does not have a good record in effective teamworking and are working to change attitudes on its projects.

It is playing a pro-active role in improving teamworking by encouraging all team members to put forward their views and ideas in open discussions.

Tesco is committed to encouraging contractors to have a greater input at the design stage in relation to buildability and value engineering. The design team is encouraged to listen, cooperate and work with contractors and take on board their advice in the interests of the project.

**To be fully effective, teams need to be fairly autonomous groups with broad terms of reference and limited supervision.
They designate the positions to be filled and the allocation of members.**

- they can provide a driving force for continually searching for performance improvements.
- they facilitate communication.
- they contain greater collective knowledge and skills and are therefore able to tackle larger issues and generate better ideas.
- they create an opportunity for greater and earlier participation.
- they are often able to turn conflict into co-operation and problem solving.
- they provide an important sense of belonging.

"Teams can solve problems and identify opportunities."

Merck Ltd, Partnership Sourcing in Action, CBI, DTI, 1996.

Keep the team together

The challenge to transform project groups into truly collaborative project teams is only the first step. The learning and other benefits acquired within maturing teams need to be captured and passed to other projects.[10]

Present procurement methods in construction work against continuity, as the Wood Report[11] highlighted over 20 years ago:

"It is arguable that contractors are more severely denied continuity by reason of current tender procedures than through uncertainties in

Ernest Ireland Construction use teamwork to continuously improve performance

Three types of teams are formed within the existing organisational structure to reinforce the idea that quality and performance improvement is an integral part of day-to-day business.

- The Quality Management Team (QMT) comprises directors and senior managers, whose task is to lead the quality improvement effort, and create the systems and environment for problem solving throughout the whole company.
- Quality Improvement Teams (QIT) are established by the QMT throughout the organisation. All staff are members of QITs. These teams aim to identify and eliminate defects and remove barriers to improvement at the operational level.
- Quality Action Teams (QAT) comprise cross functional, ad hoc focus teams to work on single issues. QATs are quickly formed, recognised and then disbanded once the problem has been solved.

This integrated strategy for change involves everyone at all levels in the organisation in a systematic and permanent process of improvement.

One tangible result of this approach is a reduction in the cost of theft and vandalism on sites from £9,000 per month to £2,000 per month.

Ernest Ireland Construction is now diffusing this approach to its suppliers and specialist contractors. This is based on joint training days and rewarding good team performance through its prestigious Subcontractor and Supplier of the Year Awards.

J. Sainsbury adopts partnering

As a regular client of the construction industry J. Sainsbury is no longer prepared to work in an adversarial manner. Such an approach is wasteful in people's time and it sustains the blame culture.

J. Sainsbury has taken on board partnering and is already seeing the benefits in cost and time savings. Alan Saunders, its Construction Resources Manager, says, "It would be easy to say that partnering is a soft option. It is not, it is hard work but the benefits will make the effort worthwhile."

total demand in their industry sector. The seeking of new tenders for each project results in a damaging lack of continuity for the individual firm. At site level, teams of men and managers built up during the progress of a job are often dispersed at the end of the contract. The advantages gained through learning particular constructional operations and sequences, understanding customer attitudes and policies, and welding together effective management and site teams are lost... Furthermore, the incentive for firms to innovate is weakened."

"Forming a new team for every project on the basis of competitive selection may reduce project costs by a few per cent, but the research findings show that the potential (of longer term relationships) for realising savings are of an order of magnitude greater than this, through a close integration of the design, procurement and construction processes."
Lynton Report[12]

Sustaining teamwork through partnering

'Trusting the Team'[10] argues that the performance of the UK construction industry can be improved through partnering, which is emerging as a better way of working together. It is a

Teams can solve problems and identify opportunities.

The Birse Transformation Programme based on improving relationships pays dividends

Birse aims to become an outstanding construction company by unlocking the collective power of its employees.

All the company's 2,000 staff have attended a two day coaching workshop designed to break down the barriers which can exist between people, even those working side by side.

The Transformation process is enabling Birse employees to enter into relationships based on open, honest and powerful communication and to reach out for what each individual and each team had previously thought impossible.

The company finds that less time and energy is now spent on activities that do not add value to the business. Empowered people are taking responsibility for their part of the business.

More powerful working relationships are being formed throughout the organisation and the strengths of teams from Board to site level are improving all the time.

Birse is now taking steps to cascade this approach throughout its network of suppliers and specialist contractors.

management approach used by two or more organisations to achieve specific joint business objectives by maximising the effectiveness of each of their resources.

Its essential basis is a free and open exchange of information to change attitudes and improve performance. It can be used on a single project (project partnering) but greater benefits flow from a long-term relationship over several years (strategic partnering).

Effective partnering in construction and in the other industries studied for this report depends on an equal relationship between partners based on trust and good communication. In particular it requires:

- choosing the right partners.
- taking an appropriate share of risks.
- devising profit sharing schemes.
- forming and agreeing procedures for resolving disputes.
- working on an open book costing basis.
- setting targets for continuous improvement.

Creating effective teams in the construction industry

Handy[13] suggests that the effectiveness of teams is determined by two main sets of factors. Firstly, 'givens' or short term constraints which can only be modified in the medium or long term. They comprise the team, the task and the environment. Secondly, 'enablers' which can be changed in the immediate present. They include leadership style, process and procedures, and motivation. This division is not always absolute and managers in construction need constantly to challenge existing constraints if they are to substantially improve performance.

The givens

Team

Size:

For maximum effectiveness and participation a team size of between 5 and 7 seems to be appropriate. For example, BAA's Genesis Project used an Execution Team comprising 7 team members.

Member characteristics:

To ensure compatibility and effective relationships it is important to ensure the best mix of individual characteristics. For example, skills, abilities, attitudes, values, beliefs, personalities etc.

Individual objectives:
To ensure trust and open communication, it is important to recognise personal agendas when agreeing common objectives.

Stage of development:
To ensure the maturity of the team time must be allowed for team development.

Task

Nature of the task:
To ensure motivation of the team and ownership of the task it is important to involve team members in determining the nature of the task to be performed.

Salience of the task:
Adding greater value to the outcomes of the task can be achieved by encouraging all team members to be creative and address more challenging tasks. For example, focusing on problem-solving and decision-taking tasks rather than routine and repetitive tasks.

Criteria for effectiveness:
Effectiveness can be better achieved when the criteria for measuring performance are negotiated and agreed with the team. For example, time for completion, cost, and quality.

Clarity of the task:
Stress, confrontation and ambiguity can be reduced through a clear statement of requirements

Environment

Norms and expectations:
The effectiveness of a team is strongly dependent upon acknowledging the differences which each member brings to the team. This is particularly relevant to project teams for design and construction where each practice or company has some norms of ways of working, style of meeting, and methods of coordinating.

Leadership position:
The effectiveness of the team is linked to the autonomy and empowerment allocated to it through its leader.

Inter-team relationships:
The performance of the team can be enhanced through better relationships between the various practices and companies represented by the team members. This is particularly significant in the context of the design and construction of projects where a number of organisations are involved.

The physical location:
The effectiveness of the team is increased by physical proximity and the removal of physical barriers. The creation of shared facilities can enhance the status and values of individual team members.

The enablers

Leadership

Leadership style:
The leadership style adopted is very important in facilitating the development of the team and its effectiveness. Leadership style should be determined by the task, the team and its stage of development and maturity.

Process and procedures

Task functions:
Effective problem-solving teams need to focus on the task in hand and adopt the following recommended process: initiating, information seeking, diagnosing, opinion-seeking, evaluating and decision-making.

Maintenance functions:
Sustaining the effectiveness of the team and the quality of relationships by encouraging, compromising, peace-keeping, clarifying, summarising and standard-setting.

Interaction pattern:
The effectiveness of the team is linked to the choice of the most appropriate pattern of interaction between the members of the team. There are three main patterns: the wheel, the circle and the web.

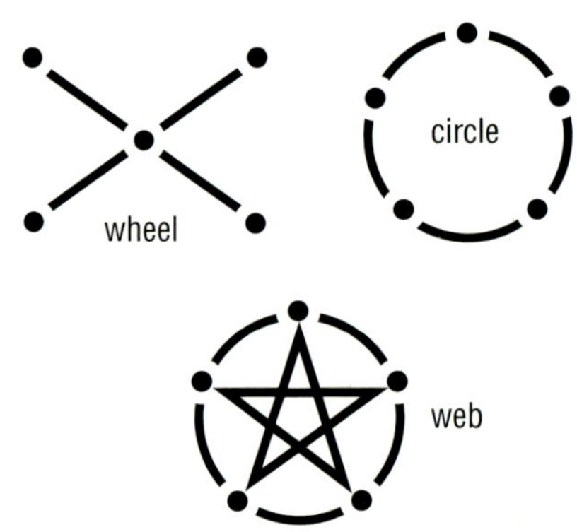

Each has advantages and disadvantages. Webs, which are more participative and good for quality of decision making in complex situations can be seen as most appropriate for the design and construction process.

Motivation:
The effectiveness of the team is dependent upon the establishment of a common cause which is sufficiently clear and salient to transcend the agendas of individual team members.

Processes are the means by which value is created for customers. They determine how work is done and how outputs are created from inputs.

A Process Approach

Essential characteristics of the process approach in the automotive industry

It is customer driven

All activities are organised and focused on a product line basis led by a product champion, with functional departments playing a secondary role

All activities are team-based and the organisation is horizontally and not vertically orientated

The whole system involves fewer actors all of whom are integrated with each other

There is a high level of information exchange between all the actors and a transparent and real cost structure

The activities are coordinated and evaluated by the flow through the team, project or plant rather than by each department meeting its plan targets in isolation

The discipline necessary for the system to function and expose problems is provided by Just in Time and Total Quality in the plant and supplier and dealer performance evaluation

Responsibility is devolved to the lowest level possible, in the plant or in the supplier's organisation

The system is based on stable production volumes but with a great deal of flexibility

Relations with employees, suppliers and dealers are based on reciprocal obligations that are the result of treating them as fixed costs.

By its very nature the design and construction process crosses the boundaries between a large number of companies and practices involved in a project. In consequence, this process is disrupted by the nature of the industry and its poorly integrated value chain.

Inter-organisational relationships are difficult to establish soundly in the face of the following regularly reported obstacles:

- fragmented nature of the design process
- inadequate design period
- late, poor and incomplete design information lacking specialist contractor input
- significant changes to the design being made during the construction phase
- poor overall planning with inadequate lead-in time
- poor consideration of specialist contractors' workload and insufficient understanding of their needs
- lack of consultation and involvement of the specialists in planning
- inadequate overall project programming and

SmithKline Beecham's emphasis on measurable objectives

For SmithKline Beecham, targets are essential to keep improvements on course, measure success and provide incentives. Graeme Telford, Global and Strategic Purchasing Manager, Capital and Construction Team, says: "In my experience people work much better when targets are set. This is equally true for the construction industry. We should be moving away from liquidated damages towards recognising and rewarding value added inputs which provide customers with the quality products they want, within the budget and programme they set."

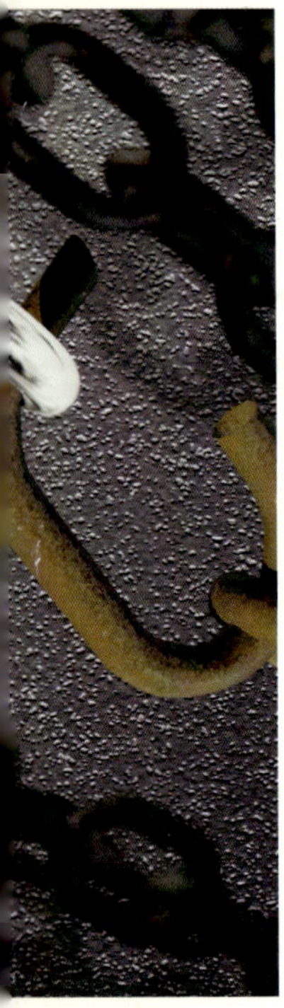

"The challenge is to harness the benefits of both approaches successfully: the short-term flexibility and efficiency associated with projects; and the continuous improvement characteristic of longer-term processes."

scheduling of work packages by the main contractor resulting in overlapping and interference

- fluctuations in demand for the products and services of the specialists
- tendering for single projects rather than a series of projects
- failure to involve specialist contractors early enough in the process

The industry is geared to flexibility and responsiveness whereas ultimate product performance may be more readily achieved by the longer-term processes associated with many other industries. The challenge is to harness the benefits of both approaches successfully: the short-term flexibility and efficiency associated with projects; and the continuous improvement characteristic of longer-term processes.

Overcoming the obstacles

Process-orientated thinking is about identifying and reducing waste and adding value. Its main objective is about enabling construction companies and practices to understand their customers' expectations meet their objectives and deliver final outcomes. The effectiveness of this approach is associated with the need to adopt a global and integrated approach which addresses all the processes together. This may be seen as incompatible with the fragmented and contractual chain which is the dominant characteristic of the construction industry.

It needs the consultant and the specialists to work in parallel throughout the design stage. Similarly, suppliers such as ductwork and controls companies must work in conjunction with the specialists. And both specialists and suppliers must be involved as early as possible in the process so that the supply chain can be thoroughly engineered and optimised.

The benefits to be derived from the process approach are greatest when it is applied to entire value chains from customer to delivery[14].

A global and integrated view is needed

Demolish the internal barriers

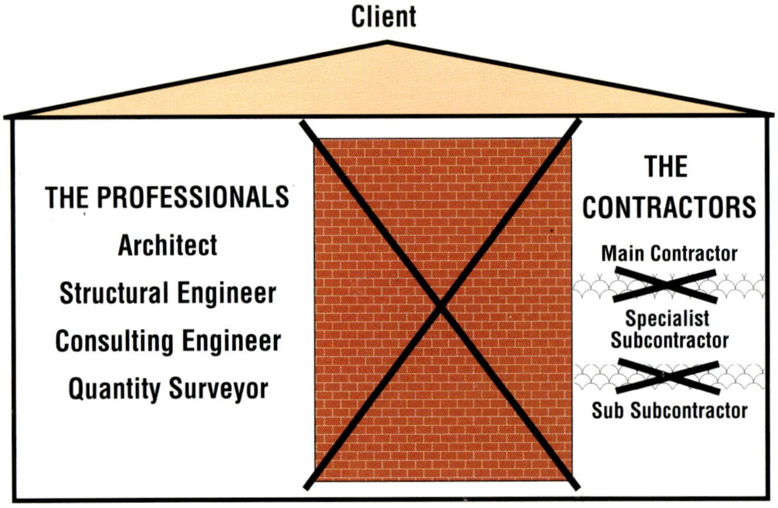

Some views from participants in the design and construction process

"The system has difficulty in using specialist contractors' ability to advantage - it is too confrontational."

"Main contractors leave ordering to beyond the last minute because they are petrified of missing the cheapest price."

"The subcontractor is the specialist but he is kept out of the design process."

"The specialist contractor is usually involved at the eleventh hour."

"The UK is the only country to use quantity surveyors who see making deductions as their main objective."

"There seems to be a deliberate policy in main contractor organisations to separate the quantity surveying aspect from the site management aspect with no overall power being given to the site manager to intercede and overrule in financial matters. The following situation regularly occurs: a specialist contractor cooperates with the main contractor's management team in out of sequence working and exceptional measures only to be faced, months later, with the main contractor's quantity surveying department which will not properly value the work and the main contractor's site management team that has no authority to intercede. The main contractor/specialist contractor operations then deteriorate causing a breakdown in relationships."

"Latham only set out to construct the team but the processes at the operational level have not seriously been considered."

The Seven Commonly Accepted Wastes in the Toyota Production System (TPS)

process design

- Overproduction
- Waiting
- Transportation
- Inappropriate Processing
- Unnecessary Inventory
- Unnecessary Motion
- Defects

which is cross functional and addresses all the processes involved in the design and construction of a project. It defies organisational boundaries. The broader the process-orientated thinking the more scope there is for reducing waste and adding value. This will help the construction industry shift from the present emphasis on cost and time to an all embracing basis of value analysis encompassing time, quality, cost and other factors as illustrated in figure 2.

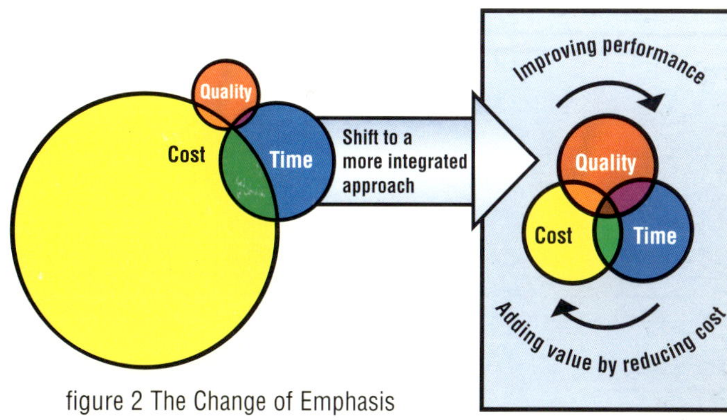

figure 2 The Change of Emphasis

Process strategies

The process-orientated approach, which started in the automotive industry in Japan, is becoming increasingly established in other sectors within the UK. Organisations in construction which are now beginning to apply it are already achieving tangible results. For example; the BAA Project Services Group believes that very substantial improvements in value are to be gained through rationalising the construction process to deliver higher levels of quality and safety.

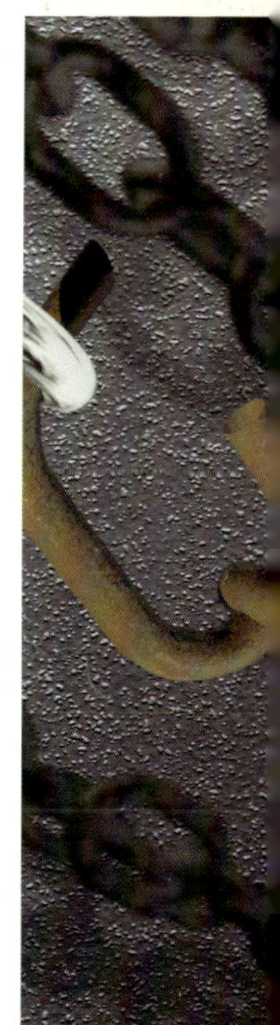

The process-orientated approach:

- describes specific inputs and outputs for each step of the process
- crosses organisational boundaries
- focuses on goals and ends rather than actions and means
- makes processes and their inputs and outputs easily comprehensible by everyone involved
- relates all processes to customers and their needs

"We had good relationships with some of the best contractors in the business but the results were not satisfactory because the process itself was wrong."

Dr. Bernard Rimmer, General Manager for Construction, Slough Estates

BAA achieves a 10% reduction in cost by adopting a process-thinking approach

In its search for cost reductions on Genesis, a project at Heathrow Airport's Cargo Centre, BAA investigated the manufacture and the delivery of every item on the project and produced intricate supply-chain maps showing every process involved.

In developing this approach, BAA examined construction industries overseas and other industries in the UK that have already faced cost-cutting exercises, and learned from other companies such as Boeing and Rover where process thinking is being used. It also held brainstorming sessions, where thinkers from construction and other industries discussed their visions of a world class construction process.

Graham Matthews[15], former Project Services Director in BAA's Group Technical Services said: "We are not reinventing wheels; we are simply taking good ideas from other industries and reapplying them to what we do here."

This approach is helped by an open-book policy with suppliers and subcontractors which gives the BAA team an insight into where there may be opportunities to share resources between the project partners and add value through avoiding duplication and overlapping. On-site costs for the project were reduced by 10%.

An underlying principle of the project is to use standardised products and processes where possible.

Although reducing the real cost of construction was the main reason behind the project, there was no random cost cutting. For instance, the BAA team accepted its contractors' profits as a fixed cost.

Improved working practices

Pursuing this broad strategy entails changes in the traditional working practices of design and construction such as:

- more complete and integrated design information through a comprehensive and clear customer brief.
- realistic project scheduling and agreement on customer-supplier relationships throughout the process.
- identification and management of the risks involved in the process.
- project partnering in order to increase understanding and develop better co-ordination of activities at all stages of the project.
- both project and strategic partnering leading to shared project and corporate objectives in order to reduce fluctuations in flow of work.
- integration of design and construction processes in order to avoid loss of value.
- early involvement of specialist contractors in the process in order to improve the accuracy of programming, buildability, and increase value and reduce cost.
- more complete and integrated design information commensurate with the needs of each project.
- sharing of learning with other project partners.
- improved exchange of information.
- better design management including sign offs and a smoother transition from the consultants to the specialist contractors, with appropiate rewards for early inputs to the project.
- greater standardisation of products and processes.
- targets and specific improvements for continuously raising the performance of specialist contractors.
- a measurement of progress.

"If you don't keep score you are only practising."
Jan Leschley, Chief Executive Officer, SmithKline Beecham

Slough Estates changes its processes to ensure early involvement of specialist contractors.

Slough Estates has developed an innovative system called PRODAC (Progressive Design and Construction) which allows the team of designers and contractors to collaborate and work together from the outset of the project.

The system is a sophisticated form of Design and Build in which Slough Estates forms a team of a construction company, designers and major specialist contractors to work together to create the best value possible. Slough Estates manages the pre-construction process during which time all are paid through the construction company. Slough also provides commercial management from its in-house team.

Specialist contractors are able to negotiate sensible margins provided they contribute their full expertise to creating value. Typical examples are M&E contractors, concrete frame contractors and curtain walling companies. In comparison with tendered consultants' design, the results have been double digit percentage savings and much improved quality and performance. Slough Estates has abolished retentions and zero defects is the aim.

Lean production

During the 1990s the automotive industry in particular has been using 'lean production' as a strategy to achieve world-class performance, which is essentially about reducing waste and adding value and is based on the effective adoption of process-orientated thinking. Lean production has been hailed as "a new way of making things which is changing the world."[16] It involves a systemic break with traditional approaches,[17] by fully exploiting the available competencies and skills in order to add greater value and satisfy customers' needs.

"Lean thinking provides more and more with less and less – less human effort, less equipment, less time, and less space – while coming closer and closer to providing customers with exactly what they want."

Womack and Jones[18]

Lean thinking is based on simple but challenging ideas to help convert waste into value. Its three main pillars are:

- management of processes and integrated logistics flow.
- management of relationships with employees, teams and suppliers.
- management of the process of change.

Toyota's lean approach to manufacturing methods has slashed development time by half, factory space by half, production time by half and defects by half.

Given the nature of construction, lean thinking can only be applied if changes are introduced in all organisations involved in the network of internal and external customers and suppliers within the design, manufacturing and assembly process. This implies better intra-as well as inter-organisational relationships based on greater collaboration. This will not only improve or enhance the quality of the service and the coordination between the different partners but will also simplify processes leading to a reduction in transaction costs, waste, overlapping and cycle time.

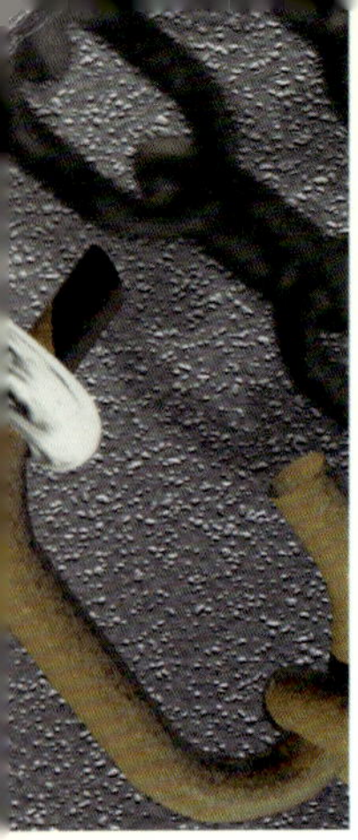

The Bovis Continuous Improvement Programme

Since 1992 Bovis has been running a continuous improvement programme which has changed and developed over time but which has maintained its focus on process improvements, customer satisfaction, learning and improving relationships within the supply chain.

The company has:

- launched its specialist contractor alliance initiative aimed at increasing communication and feedback to and from specialist contractors. This was preceded by a specialist contractor survey which sought views on how Bovis could improve.
- recently launched its new subcontract document (the White book) which aims to make contract conditions fairer.
- co-sponsored the BSRIA report on subcontractor productivity and is actioning its recommendations.
- produced a cd-rom based self development program which is issued to all staff to reinforce its values and outline basic skills. It includes a total quality module.
- embarked on a five year framework agreement with BAA and a joint venture agreement with B.P. - both of which are partnering relationships.
- since 1993, run a change foundation workshop for 500 staff and specialist contractors,

There are many current initiatives as well as completed ones over the last five years. Many have made significant improvements in the way the company and projects operate. Bovis is on a journey which progressively improves its performance but which is never over.

An approach to the adoption of lean thinking

The leading advocates of lean thinking, Womack and Jones, have identified five phases in the adoption of lean thinking:

1. Specify value.

Specifying value accurately is the critical first step in lean thinking. This can only be defined by the ultimate customer and it is only meaningful when expressed in terms which meets the customer's needs at a specific price and at a specific time. It requires effective dialogue between the customer and the supplier.

2. Identify the value stream

Identifying the entire value stream for each product/service is the next step in lean thinking. The value stream is the set of all the specific actions required from conception to delivery.

Value stream analysis will almost always show that three types of action are occurring along the value stream: steps which unambiguously create value; steps which create no value but are unavoidable; steps which create no value but are immediately unavoidable.

As value chains involve many companies, lean thinking must go beyond the firm to embrace the entire set of activities entailed in creating and producing a specific product/service. This is what Womack and Jones call the lean enterprise.

Creating such lean enterprises does require a new way to think about inter-organisational relationships. Principles for regulating behaviour between organisations, and transparency regarding all the steps taken along the value stream are seen as essential.

Tools to help identify and reduce or remove waste

Research on improving performance in other sectors of the economy has highlighted the benefits of a greater understanding of processes through the use of mapping tools. These tools are being developed to help practitioners to analyse, understand and map value-adding processes in the value stream.

These tools are also being used to identify waste and find appropriate ways for its reduction or removal.

The main benefits of a greater understanding of processes, as identified in other sectors of the economy, include cost reduction, shorter cycle times, inventory reduction, quality improvement through a better understanding of requirements, better understanding of the resources needed, clearer understanding of the role that participants play within the project process, and improved relationships between the individuals and firms involved.

It is claimed that this process-based approach can result in benefits ranging from 10% to 35% improvements in performance.

The tools which are being developed in other sectors of the economy are drawn from a variety of origins including engineering, logistics, and operations management. Hines and Rich[19] have identified the following value stream mapping tools.

Process Activity Mapping:

Involves a preliminary analysis of the process, followed by the detailed recording of all items required in each process which leads to the preparation of a map. This map can be used as the basis for further analysis and subsequent improvement through the use of techniques such as the 5W1H (asking: Why does an activity occur? Who does it? Using which equipment? Where? When? And how?)

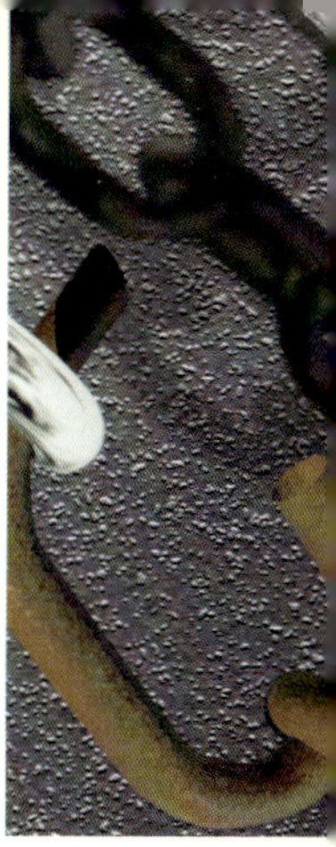

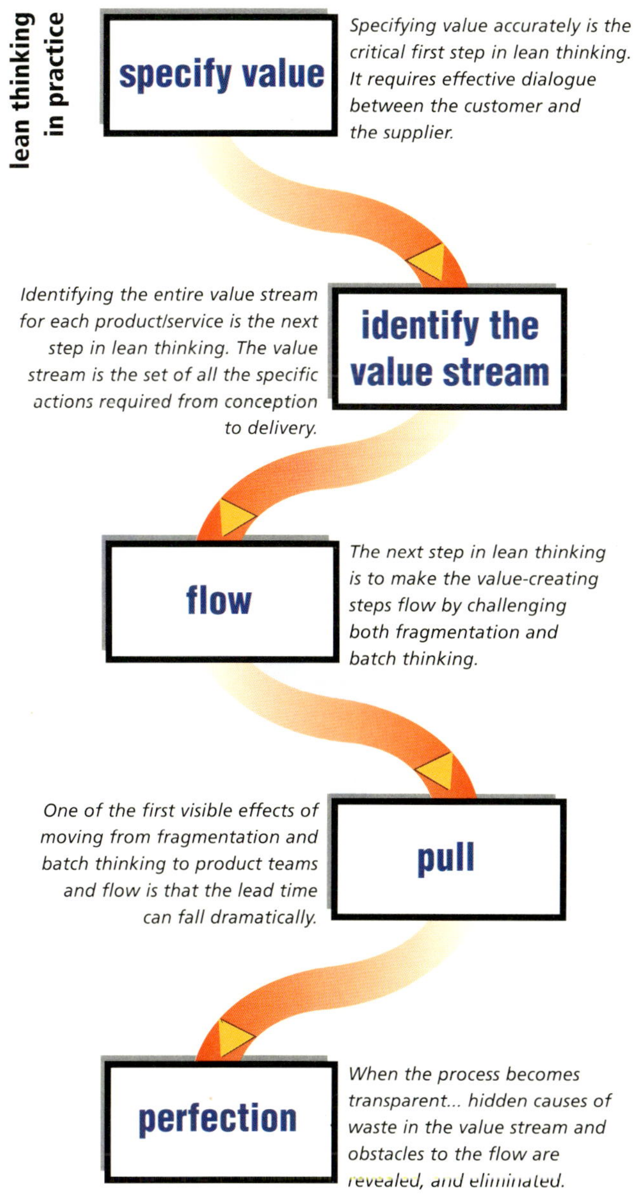

3 Flow

The next step in lean thinking is to make the value-creating steps flow by challenging both fragmentation and batch thinking. The focus must be on the product/service and its needs rather than the organisation or the resources available, so that all the activities needed to provide a product/service occur in a continuous flow. The main benefit of this approach is to challenge and redefine the work of functions, departments and firms so they can make a positive contribution to value creation.

4 Pull

One of the first visible effects of moving from fragmentation and batch thinking to product teams and flow is that the lead time (from inception/ conception to delivery) can fall dramatically. When flow is introduced, products/services requiring years to design are done in months, orders taking days to process are completed in hours, and the weeks or months of throughput time for conventional physical production are reduced to minutes or days.

5 Perfection

With successful adoption of lean thinking there is no end to the process of reducing effort, time, space, cost, and mistakes and therefore to continuously searching for improvements to achieve excellence and perfection. When the process becomes transparent and everyone involved can see everything, hidden causes of waste in the value stream and obstacles to the flow are revealed, and eliminated

Source: Womack and Jones[18]

Supply Chain Response Matrix:

Seeks to portray in a simple diagram the critical lead-time constraints for a particular process. This diagram can be used to target improvements in lead-times and inventory levels.

Production Variety Funnel:

Useful in helping to decide where to target inventory reduction and in making changes to the processing of products. It is also useful in gaining an overview of the company or supply chain being studied.

Quality Filter Mapping:

Helps to identify where defects are occurring and hence to reveal problems, inefficiencies and wasted effort. This information can then be used for subsequent improvement activity.

Demand Amplification Mapping:

Shows how demand changes along the supply chain. It provides information that can be used as the basis for decision making and further analysis and to redesign the value stream configuration and manage the fluctuations.

Decision Point Analysis:

Helps determine the decision-point in the supply chain where actual demand pull gives way to forecast-driven push. It is the point at which products stop being made according to actual demand and instead are made against forecasts alone.

Physical Structure Mapping:

Useful in understanding what a particular supply chain looks like at an overview or industry level. It helps in understanding how an industry operates and in particular in directing attention to areas that may not be receiving sufficient developmental attention.

The successful use of these tools depends on recognising and taking into consideration the context within which they are to be applied including the nature of the project, its value streams and the type of problems that exist.

Construction needs a more customer focused approach in order to provide the specialist contractors with a clearer picture of what is required of them.

Customer Focus

This strategy for improving performance has essentially come from research in the aerospace, retailing and automotive sectors. In construction, however, there is also increasing awareness of the need to have a better understanding of customers and their needs.

Excellent businesses get as close as possible to their customers, while others only pay lip service to the concept. As customers dictate product quantity, quality and service, there should be no part of the business that is closed to them. In fact, many of the excellent companies obtain their best ideas for new products and services from listening constantly and intently to their customers.

Dissatisfied customers represent major lost opportunities. It is widely acknowledged that developing new customers can cost an average of five times more than it does to retain a satisfied customer.

Who is the customer?

A customer is often defined as the purchaser who buys a commodity or service. This implies only one type of customer – the external customer who is outside the organisation and

Enhanced customer care

How other industries highlight the importance of the customer-centred approach

Recognition that there is a purchaser-supplier relationship which needs adequate time and resources to properly develop

Clarity of customer-supplier relationships and clear statements of requirements

Negotiation of goals and the mechanisms for measuring performance in relation to individual transactions, performance of the supplier-customer, and the overall value added through the relationship

Changes to the internal relationships, culture, behaviour and infrastructure within each organisation in order to facilitate appropriate inter-organisational relationships

Contracts and agreements designed to reflect the benefits of the purchaser-provider partnership and which are not simply based on a narrow range of elements such as price

High levels of trust and honesty

Openness, transparency and more sharing of information

Working towards a more equal distribution of power within relationships

Compatible values and strategies

Reliability

the end user of a company's products and services.

However, as shown in figure 3 this does not provide a broad enough view of the participants in the value chain and of their relationships. Another category of customer needs to be recognised – the internal customer.

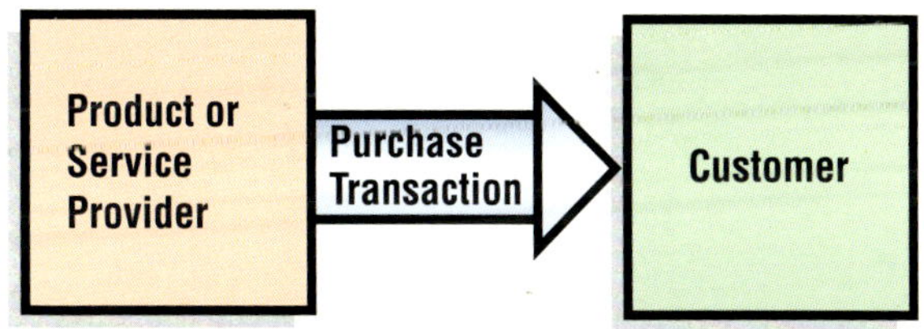

figure 3 Traditional Customer-Supplier concept
Source: Berk and Berk[20]

The internal customer is the person within the company who receives the work of another and then adds his or her contribution to the product or service before passing it on to someone else.

Customers in this sense are no longer necessarily outside the bounds of an organisation selling a commodity or service. They can be inside the organisation providing the product or service. The internal customer does not engage in trade, but instead simply receives the output of suppliers. Virtually anyone engaged in any organised endeavour has a customer.

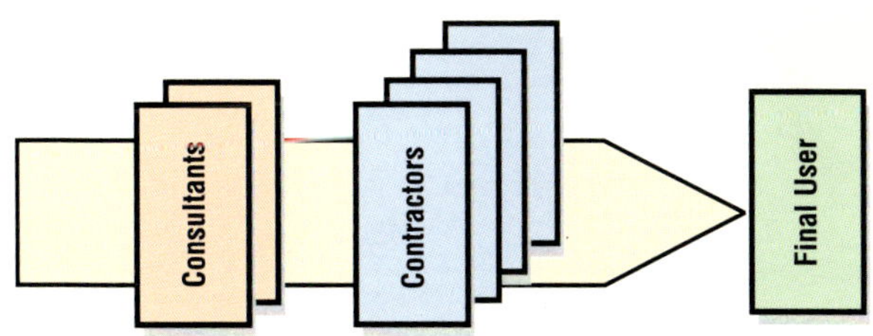

figure 4 Project participants: the concept of internal and external customers within a construction project

Customer-supplier relationships are evident in construction projects where the main contractor is the customer of the specialist contractor, and the end user or project manager is the customer of the main contractor.

Identifying the customer

Construction projects are undertaken by an amalgam of companies and practices which are independent but which must work closely together within project groups in order to

Ernest Ireland Construction's greater emphasis on its customers

Ernest Ireland Construction's successful attempt to improve its performance is based on knowing its customers and their requirements. It has undertaken regular surveys of nearly 300 of its internal and external customers to identify how it can best satisfy their requirements.

Emerging strategies within the construction industry to increase customer focus

The following approaches are beginning to be taken:

- Recognise that there are internal as well as external customers
- Concentrate on achieving customer's objectives
- Negotiate a complete and clear customer brief
- Provide incentives for teamworking
- Adopt partnering agreements
- Develop shared project objectives
- Provide payment for design input
- Involve specialist contractors early in the process
- Provide intra-company feedback
- Improve exchange of information
- Make the process known and visible

Speed of construction

succeed. When independent entities are working in a group to meet common objectives, they can be seen as internal customers within the overall project. The focus on processes, as advocated in the previous chapter, helps in identifying the customers within the project.

figure 5
Customer-Supplier chain

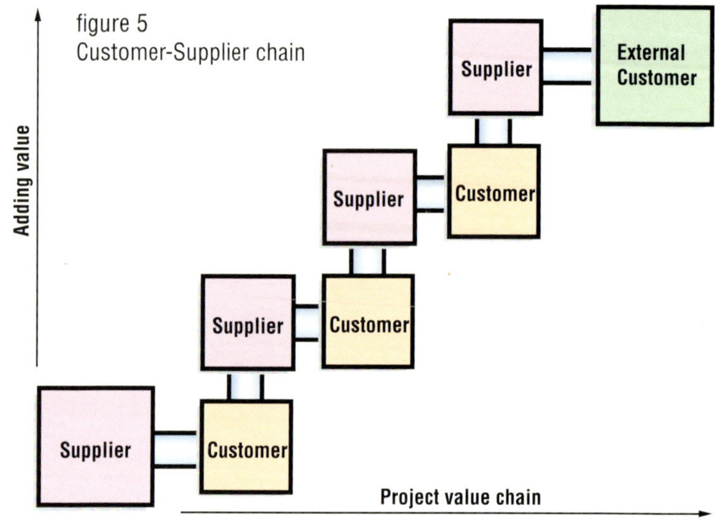

It is also necessary to understand the customer's organisation and to identify the needs and expectations of each individual within it. Many organisations in other sectors go one step further and identify the individuals within the supplying organisation who need to maintain an interface with their customer counterparts.

Specialist contractors are used to being "down the line" in the supply chain, and rarely have the opportunity of direct contact with the ultimate customer, to get to know his or her real business needs. Messages are often muddled or diluted as they are passed down the line. Discerning customers now want direct communication with the specialists whose task it is to turn design concepts into products and practical layouts; and the specialists need to learn how to listen well to the customer's needs and be forthcoming with potentially invaluable advice.

If there is no clear perception of who the customers are in the first place, then it is unlikely that there will be a clear perception of their needs either. If, on the other hand, the internal customers' requirements are agreed and met, a chain of improvements is made that reaches out to the external customer or end user, as shown in figure 5.

The customer-centred approach requires all the many parties in a process to work together as a matter of course.

Toyota uses Quality Function Deployment (QFD) to generate dramatic improvements in performance

QFD is an advanced Total Quality Management (TQM) tool for identifying customer needs and expectations. As defined by the British Quality Foundation, TQM is a customer centred philosophy of management concerned with every aspect of an organisation's activity.

Toyota used QFD to reduce the time taken to bring a new vehicle from conception into full production to the point where they were a full two years ahead of their Western rivals. In the process they also dramatically reduced the number of design modifications required, shifting the emphasis on to prevention.

QFD is a graphical tool for revealing a customer's stated and unstated needs and expectations, for making decisions in cases where these needs and expectations conflict, and for driving customer-based requirements and expectations into product development and the production process. Driven by what the customer wants, it is often described as "deploying the voice of the customer".

There are four key elements in the QFD matrix:

- Defining and agreeing the requirements
- Determining how to meet the required requirements
- Setting quantifiable goals for achieving each requirement
- Identifying and resolving conflicts in the requirements and goals

QFD matrices can also be expanded to benchmark performance against competitors

As an advanced TQM concept QFD cannot be implemented without other underlying TQM concepts being in place: recognition that most quality problems are systems related; a company culture that emphasises team working; and a willingness to invest more time in design and development in order to prevent problems downstream in the process.

Cost effective construction

Defining customer requirements

A customer's objectives need to be:

- measurable
- prioritised

The project team should then be collectively responsible for meeting – or bettering – the customer's measurable objectives in the defined order of priority. An incentive would be to share the savings or additional profits which successful teamwork achieves. Also, flexibility

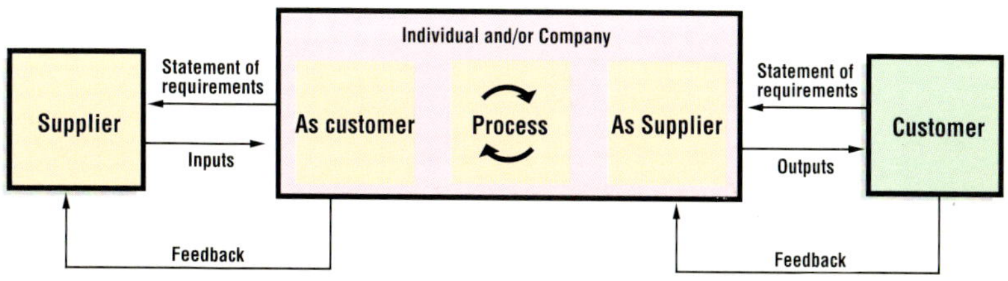

figure 6 Customer-Supplier Relationship

over the re-arrangement of fees according to input would help to ensure that if one member of the team is having difficulty, the others are willing to provide help for the overall good of the project.

A teamwork approach which is responsive to the requirements of the customer could extend into facilities management when aspects such as cost in use can be monitored in the long term. A number of consultants and specialist contractors already offer this. It could be a basis for continued collaboration after practical completion.

Agreeing customer requirements

Once the the customer and supplier have negotiated and defined the requirements, the quality of the product or service is easily determined. A defect is quite simply a failure to meet the requirements.

Key features of customer focus

Albrecht[21] suggests that customer focussed businesses are built on the following seven key characteristics:

- leadership which is mentally and behaviourally flexible.
- an understanding of and special insight into the needs of customers .
- the ability to deliver in ways that separate them from the competition.
- commitment to changing themselves in directions that are consistent with evolving customer needs and expectations.
- recognising that human energy is the greatest untapped resource in an organisation and that participation from customers, owners and employees is critical
- a caring and sharing culture.
- a drive to be best and a commitment to continuous improvement in all aspects of their organisation.

Even though the requirements are often issued by one person to another, or by one group to another, it is essential that both parties understand and agree to them, as illustrated in figure 6. If they are not mutually agreed or cannot be met they must be changed to what the supplier can deliver and the customer can accept.

There is a tendency in construction for suppliers to agree to customer requirements that cannot be met, resulting in disappointment and conflict in the customer-supplier relationship. Specialist contractors need to exercise more maturity when negotiating and agreeing outputs with their customers.

This approach encourages frequent and direct communication between customers and suppliers and can provide a basis for the development of trust and continuous improvements in quality.

Certainty of completion

Quality planning road map

Juran,[22] one of the most influential writers on quality, has suggested the following framework to enable organisations to develop a stronger customer focus.

- Determine who are your customers.
- Determine the needs of your customers.
- Translate those needs into your language.
- Develop products or services which respond to customer needs.
- Optimise the product features so as to meet your needs as well as your customer's needs.
- Develop the processes which are able to produce the product or service.
- Optimise the processes.
- Prove that the processes can produce the product or service under operating conditions.
- Transfer the process to operation.

Implementation of the Strategies

Raising Capabilities through Learning

Diffusion and Leadership

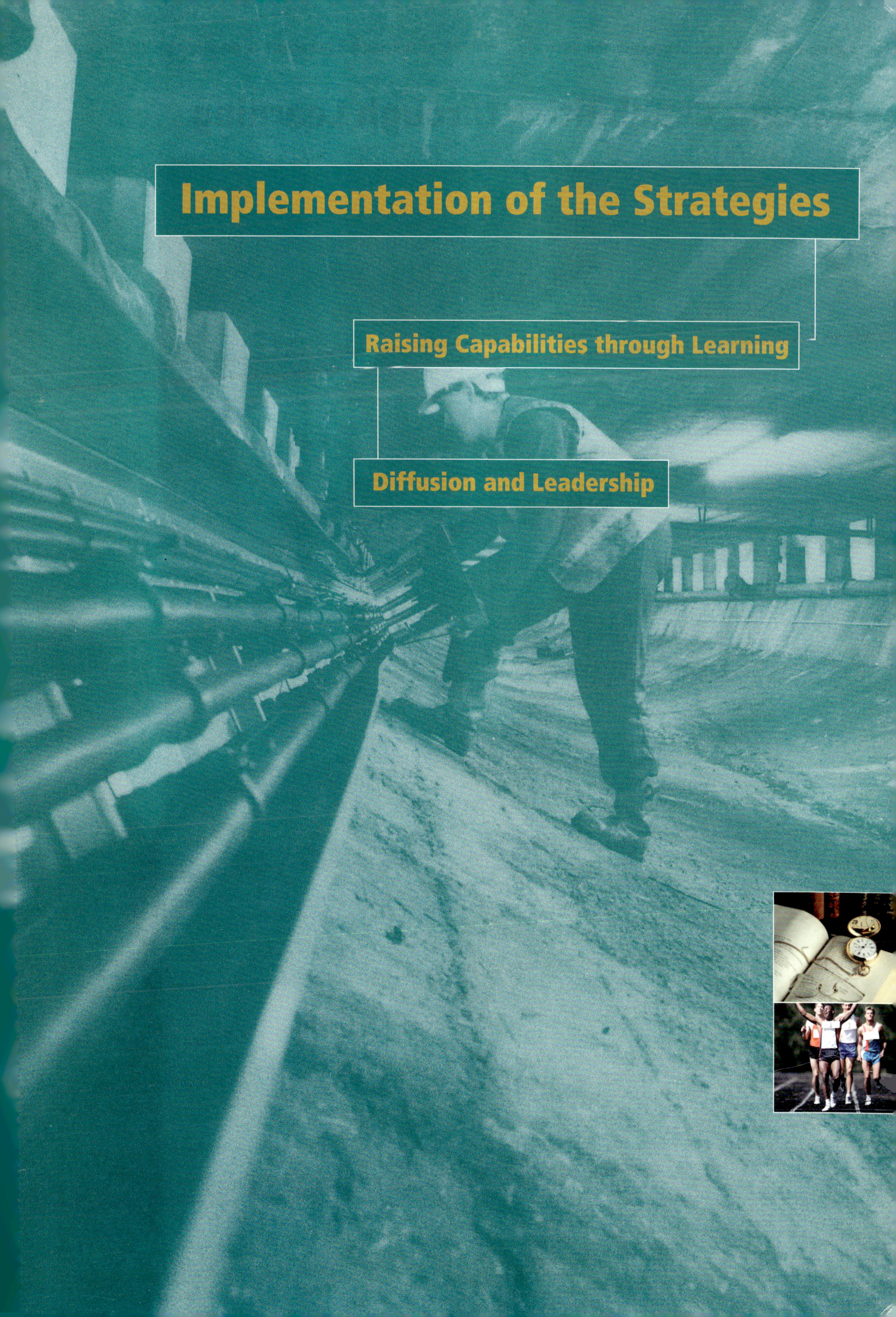

Awareness of the need to improve the internal capabilities of specialist contractors is developing very strongly.

Raising Capabilities through Learning

Construction industry views on internal capability of specialist contractors

"Specialist contractors transfer learning from project to project because of the multiplicity of main contractors and their high staff turnover. Each project seems to have a main contractor learning curve with a lack of knowledge of how to get the work properly ready for the specialist to follow through."

"Raising competence thresholds through trade associations is vital."

"Transferring learning from project to project is a good strategy but will be difficult to implement."

'Raising competence is the only way to improve quality and increase investment."

"Training should be industry led through trade associations."

The successful implementation of the three strategies advocated in this report is dependent upon greater emphasis on education, training and learning.

If specialist contractors are to play a more participative role and become truly world class, they need to place far more emphasis on learning, and to acknowledge the relationship between learning and change. The factor most likely to influence their capacity to change is their capacity to learn.

However, there is insufficient ongoing training and learning to support changes indicated in this report such as the introduction of teamwork, a process-oriented approach and greater focus on customer requirements.

Major organisational impediments to learning

The design of the organisation, specifically , the traditional hierarchical bureaucracy as a way of organising people.

The design of the job as a way of organising tasks.

Various human resources planning, control, and development systems and processes including performance appraisal and salary administration systems, and management development approaches.

Source: Cavaleri and Fearon[23]

Asking why, rather than what and how

In a seminal work 'Organisational Learning',[24] Argyris developed the concept of single-loop and double-loop learning, as shown in figure 7.

Single loop learning is where individuals and groups respond to change by detecting errors and modifying strategies within the existing norms of the organisation.

Double-loop learning, however, is more relevant to construction where there is a need to change processes and focus on the customer.

It can enable specialist contractor organisations to question the very foundations of their existence, to examine the usefulness of their values and norms. Instead of asking how they can improve what they are doing, they should begin to ask why are we doing this at all? What alternatives are there?

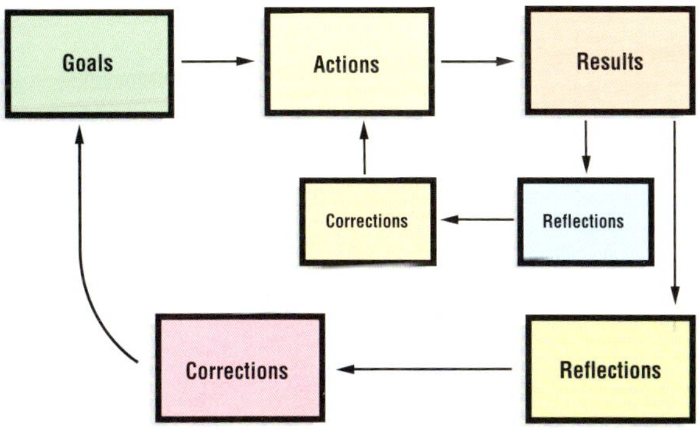

figure 7 Reconstructive or double-loop learning adapted from Argyris and Schön

Obstacles to learning for specialist contractors

Many specialist contractors are small firms in terms of turnover and asset-base, typically using craft skills and approaches.

This, and their low status in the construction process, gives rise to a number of problems which hamper the development of their capabilities:

- limited resources for training.
- tight margins.
- negligible involvement in product and process improvement.
- routine tasks with little opportunity for learning.
- shedding of directly employed labour and the rise of self employment (though this trend is changing as a result of an Inland Revenue and Contributions Agency campaign to regulate pseudo-self-employed workers).
- one-off projects.
- fragmented nature of construction.
- lack of continuity of work and short term contracts.

Preconditions of learning

- readiness: learners must have the pre-requisite knowledge, background, experience, attitudes and dispositions to learn.
- attention: performance management approaches involving mutual objective-setting and frequent performance feedback, can keep peoples' attention focused on useful activities and outcomes.
- motivation: learning is generally most productive when the learner is intrinsically motivated to learn, when it is done for the "search for excellence."

New approaches within the industry

Specialist contractors and others involved in the design and construction process are beginning to attach increasing importance to:

- learning from project to project.
- training in "soft" management and interpersonal skills.
- improving design capability.
- continuous and shared learning.
- new inter-organisational relationships.

Learning new ways of working

The Tavistock Institute and the Construction Industry Research and Information Association (CIRIA) have been working together on a DoE* funded project on Effective Learning Networks.

A broad picture is emerging of a large number of learning possibilities:

- the formal or organised network of trade associations, professional institutions and technology-based research associations
- companies working together whether in stable supply chains or in one-off projects
- informal contact between people working for different companies

*Now DETR

- improving capabilities through training in:

team building

problem-solving

programming and scheduling

Learning organisations

If specialist contractors are to be more involved and to become world class they need to become learning organisations.

"The learning organisation facilitates the learning of all its members and continually transforms itself."

Pedler, Bourgyne and Boydell[25]

The idea of the learning organisation is now a well established metaphor for management thinking in the 90s. The concept is based on the need for the organisation to fully utilize all the human resources at its disposal.

Learning must not be seen as an indulgence, but as a competitive necessity. It is not an isolated or occasional exercise but a continuous activity. It affects and involves everyone by drawing upon each individual's learning to create a whole that is greater than the sum of its parts.

"A world-class learning organisation is an organisation that achieves revolutionary rates of learning and performance improvements by becoming exceptionally skilled at creating, acquiring, and transferring knowledge, and at modifying its behaviour to reflect new knowledge and insight."

Cavaleri and Fearon[23]

However, creating learning organisations has proved difficult, for many understandable reasons. Managers can be unhappy about handing over power in relation to training. Learning requires flexibility as well as the willingness and authority to risk something new and to learn from mistakes.

Creating learning organisations involves a readiness to deal with the uncertainty and ambiguity that learning and questioning create, to accept responsibility for pursuing and creating learning opportunities, to acquire new management skills including listening, coaching and facilitating, and to develop a new culture based on trust.

In addition to focussing on what needs to be learned, the world-class learning organisation must be increasingly concerned with learning how to learn. This may well challenge the way organisations are currently learning.

"Any company that aspires to succeed in the tougher business environment of the 1990s must first resolve a basic dilemma: success in the marketplace increasingly depends on learning, yet most people don't know how to learn."

Argyris[24]

Bristol and West recognises that world-class performance can only be achieved through the efforts and development of its employees.

Ian Beveridge, Head of Human Resources Management, identifies the principal challenge: "We want to be regarded as a great place to work. The goal is to provide a challenging environment where people have the opportunity to achieve and to be rewarded for achievement. Releasing the potential that is latent in everyone is absolutely essential." ('Britain's Best Employers, Corporate Research Foundation, 1997)

Action learning

Learning by individuals as well as organisations is based on the interaction between programmed knowledge, which is the acquisition of knowledge or skills, and questioning, which is the process of exploring such knowledge in practice. Reg Revans[26] refers to this as action learning.

"The central idea of this approach to human development, at all levels, in all cultures and for all purposes, is today that of the set, or small group of comrades in adversity, striving to learn with and from each other as they confess failures and expand on victories."

Action learning must be integrated into the everyday working lives of individuals within specialist contractor organisations. It is more likely to lead to behaviour change when participants are exposed to new approaches which involve learning not only by doing, but also by group transfer of know-how and know-why.

"Collective or organisational learning is one in which people learn through working together and work together through learning."

Garratt[27]

Continuous learning is seen as central to Bristol and West's success. A broad array of training programmes enhances skills at all levels of the organisation. Staff contribute to the identification of their learning needs. From a substantial menu of courses and development activities they choose those which achieve their own objectives while contributing to the organisation's mission.

Every employee benefitted from an average of 14 days' training in 1996 – higher than the average for the industry. This is set to rise to 16 in 1997. An early indicator of the success of this commitment was attainment of the Investors in People National Standard in 1995.

Training needs are identified through a range of methods, such as skills assessment, job descriptions, performance and development review, corporate strategy and competencies. An extensive catalogue of internal and external training solutions includes individual coaching and counselling, reading material, computer-based training, video, secondments and job swops, project-based assignments, distance learning and self-study, off the job classroom activity and on the job training. Staff at each branch get together weekly as a team to ensure that they all benefit from learning and play a role in providing a world-class service. Effective learning within the organisation is promoted through recognition and rewards, including the identification of staff 'learning champions' who achieve something exceptional.

As a result of the priority given to learning, 80% of staff appointments are made from existing employees who have benefitted from a development programme.

Bristol and West recognises the value of sharing learning with other leading financial institutions. It meets regularly with competitors under the auspices of the Training Forum to exchange ideas and best practice.

Shared learning

The design and construction process involves a large number of independent but inter-related organisations, and improvements to the project process require learning across organisational boundaries. Learning must be shared by all participants: the individual organisation, the project teams, supply chain alliances and trade associations.

Life-long learning

World-class organisations are increasingly concerned with the accelerating rate of change within the business environment and the need for continuous improvement.

As they face greater customer- and process-orientated approaches they must develop their people to take on a more challenging range of responsibilities. This is also why world-class organisations are promoting the need for both deep and life-long learning in order to sustain competitiveness.

The Construction Industry Board's Working Group 9 in its report "Educating The Professional Team" recognises that life-long learning is an essential characteristic of a practising professional.

Interdisciplinary learning

Training the next generation of entrants to the

The Construction Productivity Network (CPN)

CPN provides a link between companies in all parts of the construction industry and professional, academic and research bodies, with the common aim of improving productivity in UK construction.

The network:

- enables sharing of the benefits of research, experience and innovation
- creates opportunities for the practical application of available knowledge
- promotes the exchange of expertise between large and small organisations

The network exchanges, informs and improves through workshops, visits to research centres and sites, an annual conference, and a range of publications.

construction industry must emphasise an approach to learning which breaks down the 'divide' separating professionals from the specialist contractors. Status within the industry should be earned on merit.

Learning that is pragmatic, which foresees and overcomes obstacles and hence improves performance, should be seen equally as meritorious and indispensable as other more conceptual forms of learning.

A career within specialist engineering would hold greater attractions for the young if their education and training reflected the fact that they will be working in collaboration with architects, planners and other disciplines. Courses that have interdisciplinary modules, for example, could be supported by preferential government grants.

Education and training for careers in the industry must also cultivate the teamwork ethic, listening skills and communication. All participants in the construction process should be given opportunities throughout their careers to learn more about each other's roles, and to understand how the difficulties they face can be resolved through cooperation and mutual support.

There is a case for developing a new breed of Architect/Engineer, as exists in Germany, Sweden and Japan; this would remove the barriers and difficulties of communication between architects and engineers, and should also redress the imbalance in the popularity of the degree courses for architects and engineers in the eyes of the young and their parents.

The key skills associated with world-class organisations

Multifunction instead of single function.

Short skills life cycle rather than long skills life cycle.

Skill life shorter than employee working life.

Emphasis on flexibility and creative problem solving.

Need for understanding of the whole system.

Variable skill relationships rather than stable skill relationships.

Cross trading.

Multiskilling.

Continuous education.

Source: Bessant[28]

Rover increases competitiveness through learning

In 1976 Rover's predecessor British Leyland was struggling. It had the worst industrial relations in the European motor industry. Quality was poor, deliveries erratic and customer service appalling. In order to recover, the company had to learn new ways quickly. A fresh start was made by appointing new leaders and by giving managers the authority to manage.

Rover's 16-year partnership with Honda from 1978 presented further opportunities for learning. During that time six models were jointly designed and built and both organisations learned from each other's different production development processes.

Even greater emphasis was placed on learning when British Aerospace acquired Rover in 1988. By 1994, the major advances Rover had made, as well as its unique and successful experience of partnership with Honda, was reflected in the £800 million paid for the company by BMW – over five times more than British Aerospace had paid six years earlier.

Rover stress the importance of leadership by top management, who must create the conditions for growth, change, quality and learning. A vital role is also played by 'change agents' – individuals who are prepared to cross barriers, attack difficulties and push teams into new areas. They are instrumental in transferring and implementing learning, not just within Rover but also within partner organisations such as Birse.

Creating a learning environment is of little value unless it results in things being done better. This has been achieved through action learning, or learning on the job, and the establishment of a system for checking that progress is being made.

Through its partnerships with Honda and BMW, Rover has benefitted from using the best practices from abroad. It is involved in international networks including the European Consortium for Learning Organisations, the European Life-Long Learning Initiative and the World Initiative for Life-Long Learning.

Rover's vision remains the same as it was in 1989 – to be able to produce better quality cars more quickly and for less money. But learning is now seen as being even more important in achieving these goals.

The overall returns from the company's growing commitment to learning are of far greater value than the effort invested.

Source: 'Improvement Via Leaming Organisations', by Barrie Oxtoby, Rover Group Cars, in the Conference Proceedings of the Construction Productivity Network Annual Conference, 'Improving Competitiveness From Abroad', BRE Garston, 1996

Learning - the future

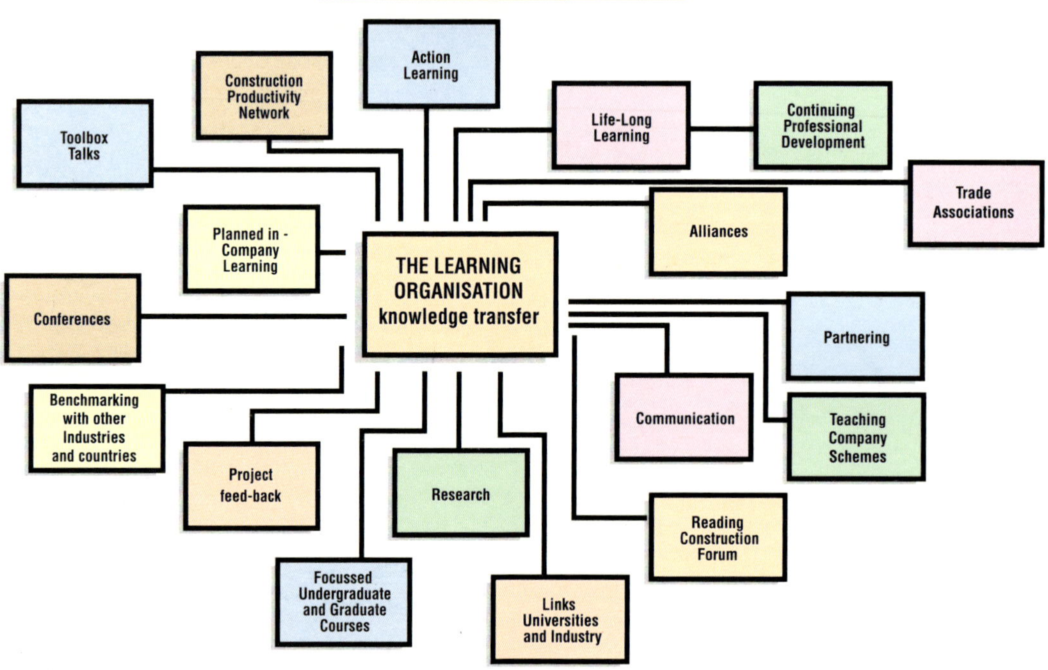

A steady accumulation of manageable improvements will eventually transform the culture and working practices of the construction industry.

Diffusion and Leadership

Industry views on successful implementation

"It is very important that specialist contractors are helped and supported by the other participants in the process."

"Customers are taking more interest and seeking fairer treatment for specialist contractors."

"We should abolish the term subcontractor as it encourages the widespread perception of 'subbies' as substandard, subordinate and submissive. We should raise the status of specialist contractors as they do most of the work."

"We need to stop the current practice of 'subbie bashing'. This needs approaching from the professional organisations with severe penalties being imposed for non-compliance with contract law."

"Collaboration and trust must be customer led."

"The customer should always lead."

"Those who can, lead, those who can't, follow."

"Each party should have a clear leader for the project."

Given the complexity of the design and construction process and its project-based character, the appropriate starting point for improvement is at the project level.

The project is the most readily available context within which specialist contractors can be influenced and supported in their attempts to improve their performance.

Experiences from other sectors show that the shop floor, or operational level, is a successful starting point. In the construction industry, the project corresponds to the shop floor.

Research in construction and other industries also suggests that change is most effectively implemented when a lead partner in a project champions improvement.

Currently, this lead role is often provided by customers and main contractors, but increasingly other participants, including specialist contractors, are exercising leadership in improving the construction industry's performance.

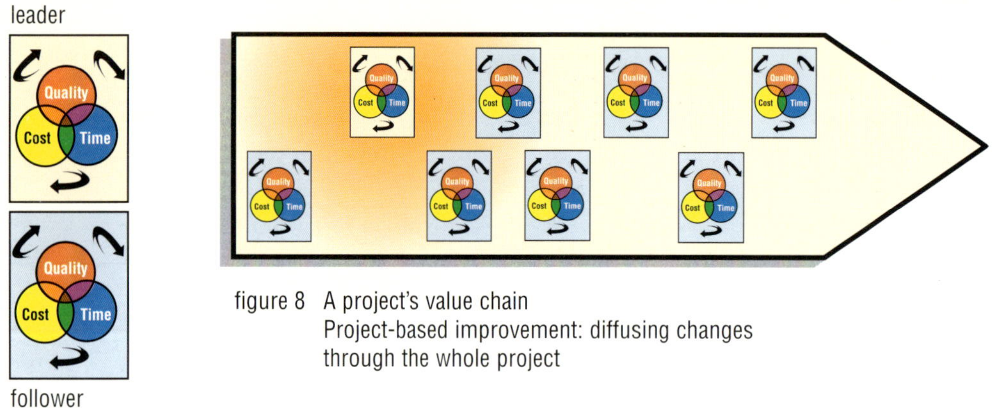

figure 8 A project's value chain
Project-based improvement: diffusing changes through the whole project

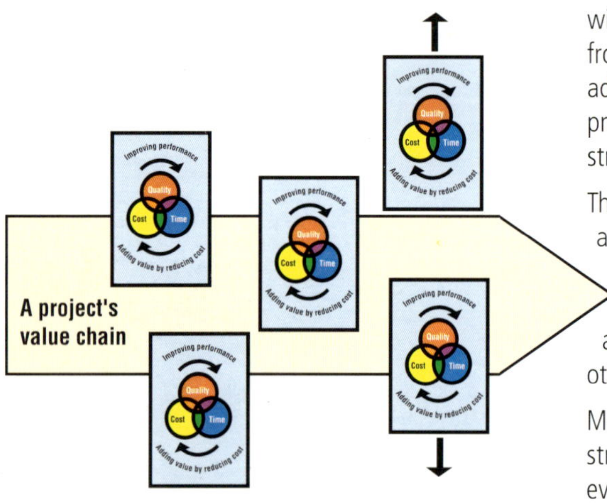

figure 9 Company-based improvement: capitalising on project learning

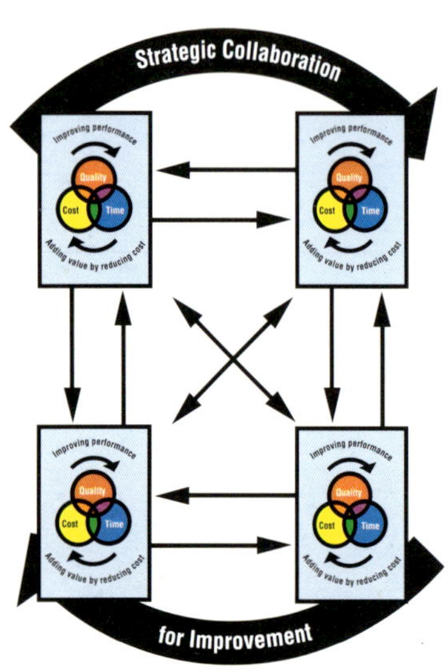

figure 10 Strategic collaboration for more significant and sustained improvement

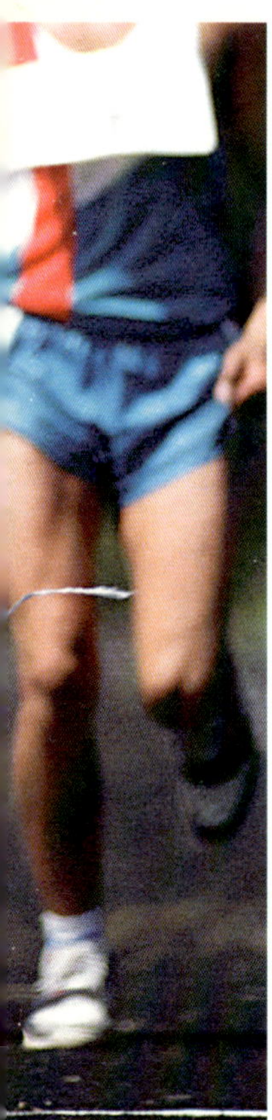

Diffusing change

The process, as illustrated in figure 8, can start with the specialists transferring best practice from the project to their own company. They can adopt changes learned within the project to improve their performance by capitalising on their strengths whilst addressing their weaknesses.

The next logical action is to extend learning and best practice beyond the boundaries of projects and companies, as shown in figure 9. This dissemination can take place as specialist contractors become involved in other projects.

More mature and richer relationships based on strategic partnering and alliances will lead to an even more significant and sustained improvement in performance, as shown in figure 10. In other industries partnering has resulted in dramatic gains in added value and reduced waste which are the basis of lean supply. It has also created greater equality and trust between partners who work together in the longer term to fully satisfy customer needs and improve competitiveness.

Having begun to implement change within clusters of companies and projects, the next challenge is to diffuse best practice throughout the whole construction industry, as shown in figure 11.

Other industries build relationships

National strategies for the development of world-class suppliers are currently being developed and implemented in the automotive and aerospace sectors.

The Society of Motor Manufacturers and Traders (SMMT) Industry Forum was launched in April 1994 to help improve the international competitiveness of the UK's automotive components suppliers. It has developed guidelines to accelerate performance improvement to world-class level.

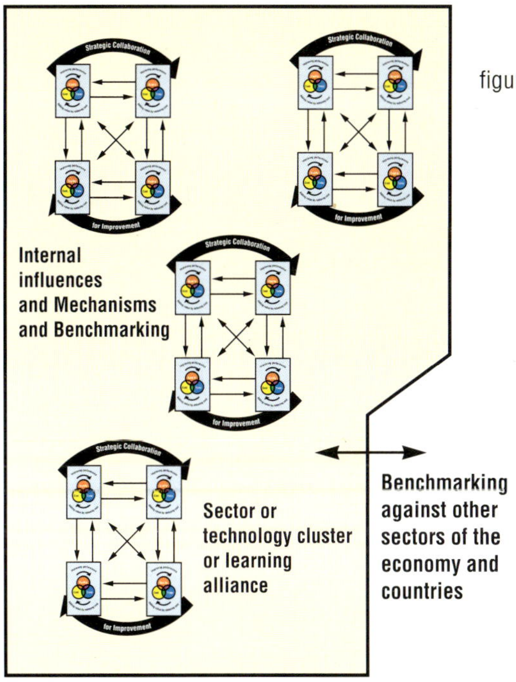

figure 11 Sector-based continuous improvement: raising the competitiveness of the whole industry.

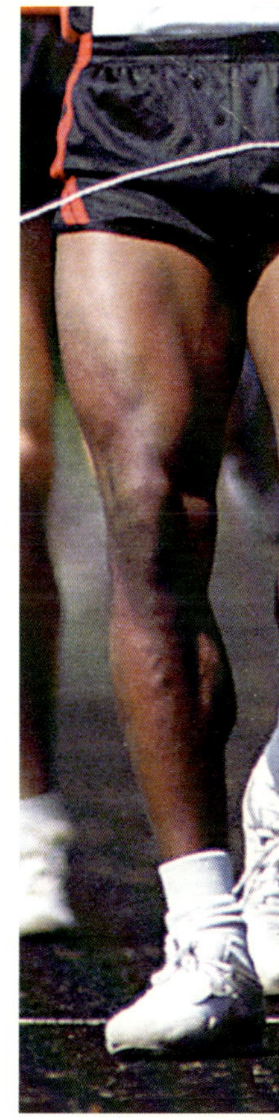

It advocates six components of success:

- process improvement.
- new product development.
- teamworking.
- problem solving.
- policy and strategy deployment.
- supplier development.

The Forum's plans include:

- the creation of a national supplier development programme which involves the recruitment of leading experts from around the world to train British counterparts in best practice supplier development techniques.
- the introduction and adoption of a measurement system which will allow a company to assess its performance and that of its customers and suppliers from its perspective.
- the coordination on a national basis of the many widespread regional and other initiatives to improve performance that are currently underway throughout the motor industry.
- the establishment of closer links with academics, particularly through the 'virtual university', to ensure that skills throughout the industry are right for the future.

Within construction, similar steps are being taken. The Roofing Industry Alliance aims to give a lead in raising standards among its members. Roofing is one of the largest specialist sectors of the construction industry, turning over £2bn and employing around 44,000. The close relationship that exists between its contracting and manufacturing sides provides favourable conditions for introducing new collaborative approaches to greater competitiveness and customer satisfaction.

The Society of British Aerospace Companies' Code of Practice: 'Working Together'

'Working Together' acknowledges that the industry's competitiveness depends largely on the performance of its supply chain. No single company can act independently of its customers or suppliers.

In the last 10 years, the UK share of the world market for aerospace products has fallen from 13 per cent to 9 per cent with new entrants and existing competitors presenting new challenges, and customers expecting both reduced costs and increased satisfaction.

The Code has been in existence for 18 months and over 80% of the Society's members are now signed up. It's early days yet but initial feedback is positive. The Code is welcomed and the general view is that it is beginning to change the culture of the industry. Strongest support comes from the small and medium-sized enterprises in the supply chain.

The six main tenets of 'Working Together':

- communication
- design for manufacture
- continuous improvement
- supplier qualification and auditing
- commercial agreements
- ethics

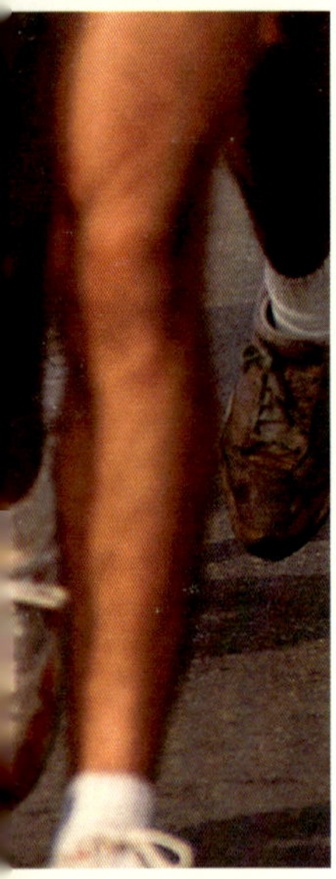

The Roofing Industry Alliance improves competitiveness and client satisfaction

The RIA was formed in January 1997 to improve competitiveness and customer satisfaction and contribute to the 30 percent cost saving called for in the Latham report.

Three working groups are investigating key issues.

The customers' working group is asking property owners, lenders, insurers, the professionals and main contractors what they want from the roofing industry.

The second group is reviewing accreditation and certification, particularly of roofing contractors, although manufacturers, merchants and stockists will be heavily involved, in an attempt to drive out the unprofessional element.

The third group is looking at staff recruitment and training.

Project based improvement: BAA's Genesis Project

The Genesis project was used as a test bed for new ways of working which aim to meet BAA Chairman Sir John Egan's target of a 50% cut in construction costs. In this project, BAA assembled a core team of individuals who had the experience, vision, enthusiasm, drive and above all the open-mindedness to change not only themselves and their companies but even the way in which the construction industry operates.

Key features of the project included the emphasis on teamworking, the earlier involvement of specialist contractors and the reduction of cost and time.[15]

The role of supplier associations in implementing change

At the beginning of the lean transformation the lead firm teaches its suppliers how to follow their examples through one-to-one supply development. This can usefully be supplemented by the formation of a supplier association to encourage shared learning among suppliers and accelerate the change process.

Professor Peter Hines, co-director of the Lean Enterprise Research Centre at the Cardiff Business School, together with the Welsh Development Agency, has been instrumental in bringing the supplier association concept to the UK. There are now some 25 supplier associations in the UK involving over 350 companies.

Leaders and champions

The successful implementation of the three strategies proposed in this report depends upon key participants championing change.

They can bring several types of influence or power to bear in order to get things done:

- influence to reward
- influence to punish
- power of authority
- influence of expertise
- influence of character
- power to empower
- power of reorganisation

"There is a rule of thumb which says employees judge a manager's commitment to any change ten per cent on what he or she says and ninety percent on what he or she does."

Binney and Williams[29]

Figure 12 provides a framework for identifying the degree of interest and influence that each participant in the design and construction process has in improving the performance of specialist contractors. Participants 1 to 3 have no interest or no influence in promoting change. The key player, the one who not only can promote change but also wants to, is Participant 4.

Customers, as well as construction managers and main contractors, are becoming a significant force for change as they seek to drive down costs and add value. Wates Construction Ltd, for example, is currently undertaking a major national survey of its relationships with

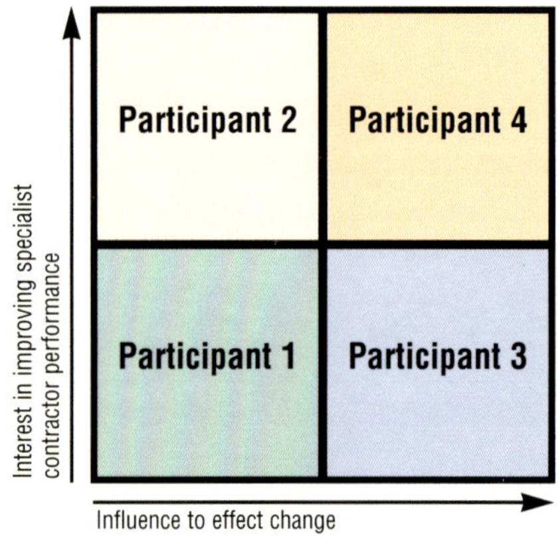

figure 12 Categories of commitment to the implementation of change

Core elements of partnership in raising the performance of suppliers in the UK's automotive industry

The Society of Motor Manufacturers and Traders (SMMT) Industry Forum has developed the following principles on which to build relationships:

- consistent understanding and adoption of the philosophy of partnership by all functions in all companies in the customer/supplier relationship
- management of inter-organisational relationships
- commitment to continuous improvement and shared benefits
- open exchange of relevant information
- complete understanding of the real costs to both parties
- establishment of common objectives focussed on customer needs
- establishment of world-class targets through benchmarking and clear performance evaluation
- agreement on a checklist of criteria to evaluate the capability and performance of a supplier
- accurate forward forecasts and continual dialogue to update/amend
- early involvement in new projects
- combined resources to tackle problems

Irvine Whitlock benefits from Construction Managers championing change

Having experienced most procurement routes Irvine Whitlock, a specialist building contractor, welcomes the opportunities to become part of the project team that effective Construction Management can offer.

Construction Managers often involve them at an earlier stage in the construction process allowing them to contribute to providing more accurate budget costing, improving buildability, adopting value engineering, and providing more accurate programming. Irvine Whitlock considers this greater involvement to be extremely beneficial for the morale of specialist contractors.

specialist contractors. This survey has identified the main problems and barriers that are preventing the specialist contractors from adding greater value for the customer. In collaboration with their specialist contractors, Wates are now developing strategies to provide their specialists with a more participative role across a number of key areas for improvement including quality, health and safety, design and specialist selection, development and rewards.

There is a role for client-industry groups such as the Construction Industry Board (CIB) which is implementing Sir Michael Latham's `Constructing the Team' review of the construction industry. Its major tasks are to implement the Codes of Practice now produced, bring forward tomorrow's working methods and steer relevant goverment research.

Government also has a role to play in encouraging change. It is already jointly involved with the industry in the CIB. Its collaboration with the automotive industry goes even further. A competitive industry selling more cars all over the world needs supply companies that are world class. A two-year, DTI sponsored 'Learning From Japan' initiative has helped identify the action that must be taken to create a more successful and vibrant automotive industry in the UK. The DTI has recruited secondees from Nissan, Rover, Toyota, Honda and Ford who are ensuring that the help given by the DTI will be practical and relevant. It has also opened an office in Birmingham with a national remit to help raise the performance of the automotive industry.

The power to empower

The exercise of power is frequently necessary in order to make progress towards goals held in common by leaders and followers - or to enable the views to become common.

The challenge of the leader/champion is not to drive change but to release the potential of specialist contractors and encourage them in the direction of improved performance. It means a planned development of their ability to make and implement decisions in an ever-widening range of situations – in short, to empower them. To develop people this way takes firm and resolute leadership.

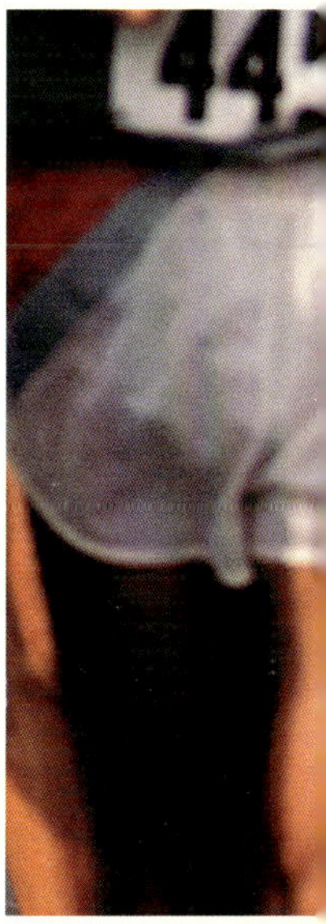

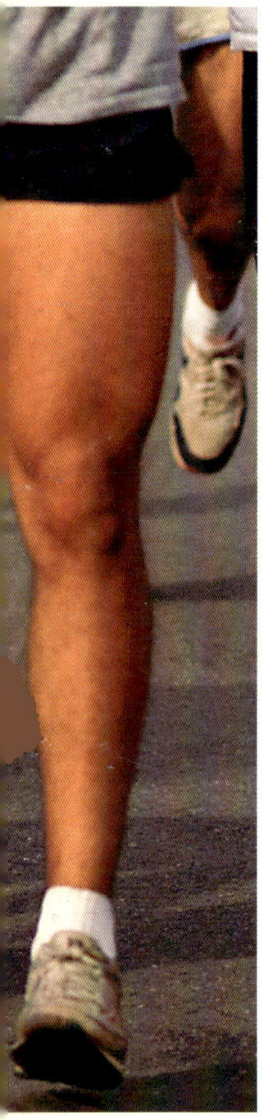

Somerfield's strategic collaboration with suppliers

At a time when Somerfield's performance was very poor and relationships with suppliers were very adversarial, the company viewed the introduction of partnering as a matter of survival in the increasingly global and competitive retailing market. It was seen as the way to improve Somerfield's overall performance and the total quality of the service it offered its customers.

So far long term strategic alliances have been developed with 20 large and well-known organisations.

Partnering is used as a tool to be more customer-oriented and to measure the performance of the whole supply chain. The outcome of performance measurement is revealed and made visible to everybody and becomes a powerful tool for identifying problems which need to be investigated in order to help partners improve their performance. A best supplier of the year award motivates suppliers to continuously search for improvement.

A major objective is to identify best practice through joint planning and communication. Teams from both partners are regularly brought together to ensure that relevant decisions are made quickly and when necessary to solve supply problems and improve the process.

Both partners agree targets such as business growth, volume of sales, market share, service level, stock level and forecast accuracy. They also hold brainstorming workshops to generate fresh thinking. The search for new ideas is reinforced by the 'job insight' exchange of staff. Visits are organised to understand each other's needs and requirements.

Somerfield values the partnership programme for the commitment and loyalty it has generated in its suppliers.

The factors necessary for a successful change programme

- Champion who embodies the new vision.
- Clear and sustained goal to which people can commit.
- Clarity of purpose for change.
- Recognition of the resistance to change.
- Being forthright.
- Effective communication.
- Proportionate allocation of responsibility.

- Recognition of the limitations of empowerment and the need for leadership.
- Development of effective teams and leaders.
- Change of structures to help develop a new culture.
- Creation of winners to sustain motivation and competitiveness.
- Actions for fast and early successes to increase momentum.
- Compassion for the losers in the process.
- Acceptance that unintended consequences may occur and action to minimise their impact.

Source: Eccles[32]

Tom Lambert[30] has identified the requirements of leaders who can empower. They should:

- retain a full understanding of what is happening
- set a clear, unambiguous direction for their teams and ensure they remain on course
- offer support and clear the way for action without taking over from those delegated for the task
- continuously assess performance, reward progress, and support individual and team development
- build trust through shared success and share information and knowledge whenever it is possible to do so

"Today's leaders understand that you have to give up control to get results - they act as coaches not as 'the boss'."

Waterman[31]

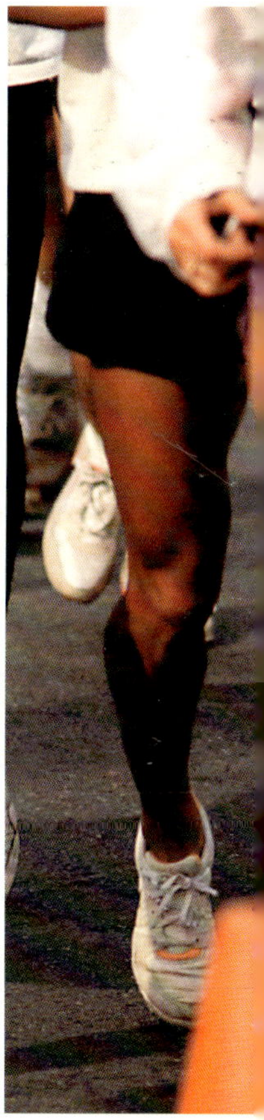

Q Flooring Systems (QFS) benefit from customers championing change

A number of customers, including Asda and Rover, specify/nominate Q Flooring Systems Ltd for their developments. These long term arrangements have led to deeper and more trusting relationships through which QFS feel more valued and an integral part of the development team.

This means that QFS can be much more focussed on their customers' needs and has more opportunities to suggest ways of reducing costs and adding value.

Also, there are benefits for QFS. With customers offering packages of work they are able to plan their workload in a more strategic manner.

QFS suggest that similar partnering arrangements with main contractors and other champions of change should also be developed.

Taking the right steps towards change

Careful planning and preparation are needed if the implementation of change is to be successful. Kotter[33] proposes eight steps to transforming a company.

1 Establishing a sense of urgency

Examining market and competitive realities

Identifying and discussing crises, potential crises or major opportunities

2 Forming a powerful guiding coalition

Assembling a group with enough power to lead the change effort

Encouraging the group to work together as a team

3 Creating a vision

Creating a vision to help direct the change effort

Developing strategies for achieving that vision

4 Communicating the vision

Using every vehicle possible to communicate the new vision and strategies

Teaching new behaviours by the example of the guiding coalition

5 Empowering others to act on the vision

Getting rid of obstacles and resistance to change

Changing systems or structures that seriously undermine the vision

Encouraging risk taking and non-traditional ideas, activities and actions

6 Planning for creating short-term wins

Planning for viable performance improvements

Creating those improvements

Recognising and rewarding those involved in the improvements

7 Consolidating improvements and producing still more change

Using increased credibility to change systems, structures, and policies that don't fit the vision

Hiring, promoting and developing people who can implement the vision

Reinvigorating the process with new projects, teams, and change agents

8 Institutionalising the new approaches

Articulating the connections between the new behaviour and corporate and project success

Developing the means to ensure leadership development and succession

Action Plan

In order to ensure that this report does not remain "on the shelf" but becomes of practical use to everyone in the team we have created an action plan which will, if followed, start to move us in thr right direction. The action plan is split into 4 sections:

Improving Relationships

Improving Processes

Improving Customer Focus

Improving Learning and Knowledge Transfer

Steps towards world-class performance

The actions outlined will help you to move towards world-class performance. They follow from either the conclusions within this report or the related surveys or other initiatives of the Reading Construction Forum. Some can be started within your company, some need other participants to join with you, and some require further action at industry level.

The main barriers you will have to overcome are:-

- fragmentation
- transient and often adversarial relationships
- insufficiently understood processes

You cannot do everything at once and you will have your own priorities. Start with one or two things you can achieve, and then build on them. For example, why not write to key members of your supply chain, tell them what you are trying to do and ask if they will commit to helping you.

GROUP A - Actions you can take within your own company

1. Develop a communications plan within your company which aims dramatically to improve communications upward, downward, sideways and to and from your supply chain. Agree it at Board level and sign everyone in the company up to it.
2. Get board's agreement to promoting "openbook" costing on all projects you are involved in implementing.
3. Agree with the board that you must create a "no blame culture" to release people's interest and ideas. Publicise this and work out how to implement it.
4. Allow time and resources for "team building" within your company.

   Complete a simple checklist of good team building and monitor and publicise.
5. Set up post contract reviews and ensure that everyone learns from them. Publicise and praise successful outcomes that showed improvement.
6. Hold regular communications meetings with staff on topics such as process improvement and knowledge transfer.
7. Ensure your front line staff are committed to forming relationships with your customers, consultants and suppliers.
8. Take a personal responsibility to promote and sell this report to as wide an audience as you can.
9. Revist your company's internal marketing system and ensure you are getting good messages across.

GROUP B - Actions customers and other partners can take to make Specialist Contractors more effective

1. Be prepared to use a wider range of criteria than just lowest price in selecting specialist contractors.
2. Invite specialist contractors to participate in the early stages of project development
3. Treat specialist contractors as equals in the team.
4. Consider long term relationships with partners
5. Promote a no blame culture
6. Adopt open book costing
7. Develop a culture of trust with your partners
8. Phase out retention and phase in zero defects.
9. Share profits from cost savings to reward appropriate involvement and develop new attitudes and incentives for teamworking.

GROUP C - Actions you can take which will need input and help from your customers and other partners

1. Communicate constantly and deeply with your partners, including customers, consultants and suppliers. Invite them to visit and discuss issues of common interest such as the "process", teamworking and incentives.
2. Agree with your customers that open book costing will be adopted on their projects and that the same approach will be applied to your suppliers.
3. Persuade your customers and their advisors to abandon adversarial forms of contract in favour of procurement terms which align everyone's interests.
4. Convince your customers to resolve disputes by mediation.
5. Agree with your customers that a "no blame" culture is more productive and should be adopted.
6. Agree fair and clear payment terms. Pay suppliers on time and expect customers to do the same.
7. Work with customers and advisors to phase out retention, and apply a zero defects policy.
8. Revisit the number of customers and other partners with whom you have long term relationships.
   Use these relationships to help you become more effective.

"The starting point is to recognise that Specialist Contractors are essential to the success of a project and that they have much more to offer to all partners than is realised at the moment."

GROUP A - Actions you can take within your own company

1. Focus on process management in order to deliver greater value and reduce waste.
2. Identify your key processes.
3. Place your customers at the centre of your process thinking.
4. Develop process maps for your project involvement and start improving.
5. Work to simplify and increase standardisation of your products, services and information.

GROUP B - Actions customers and other partners can take to make Specialist Contractors more effective

1. Involve specialist contractors early in the development process for a project (day one) in order to add value and reduce waste.
2. Recognise that specialist contractors are at the centre of the design and construction process.

GROUP C - Actions you can take which will need input and help from your customers and other partners

1. Work to make the whole process known and visible so that the process inputs and outputs and their costs are easily comprehensible and measureable by every-one involved.
2. Learn to use simple tools, such as flow diagrams, to describe your business and project processes in order to:
   - measure your present performance
   - know who is doing what, where, when, why and how
   - rationalise information and optimise its flow
   - clarify the expectations, roles and responsibilities of each participant
   - agree payment terms based on processes
   - simplify processes
   - identify, agree and target areas for short, medium, and long-term improvement
   - identify the skills and competencies needed for improvement
   - identify where costs are being incurred and value added
   - investigate how you can link your processes with those of other participants
3. Re evaluate objectives and revise methods in the light of developing experiences.
4. Identify, manage and allocate appropriately the risks involved in the process.
5. Work to increase standardisation of products, services and information.
6. Move the emphasis from commercial to technical considerations

GROUP A - Actions you can take within your own company

1. Identify your internal customers and start improving your focus on them and their needs.
2. Develop a stronger customer focus so that all your processes relate to your internal and external customers and their needs.
3. Seek to become a responsible and mature customer and supplier.
4. Start a customer care/customer focus programme.
5. Be aware that customers' needs and aspirations are constantly changing and rising. The only way to be in tune is constantly to "ask the customer"
6. Arrange regular reviews on performance with your main customers at project and company level.

GROUP B - Actions customers and other partners can take to make Specialist Contractors more effective

1. Make the specialist contractor aware of the end user's requirements
2. Agree measures of performance and priorities with Specialist Contractors.
3. Treat suppliers and specialist contractors as if they were customers in order to unlock their potential for the benefit of the project.
4. Involve end user groups in development and design planning along with specialist contractors who supply or make the end user equipment.
5. Provide incentives for Specialist Contractors to achieve or exceed project objectives.

GROUP C - Actions you can take which will need input and help from your customers and other partners

1. Adopt the attitude that every member of a project is a customer as well as a supplier and each has their own needs and contributions to make.
2. Identify customer-supplier relationships throughout the whole process.
3. Develop clear customer-supplier relationships by negotiating a complete brief and statement of requirements.
4. Recognise the importance of your internal as well as external customers in the project process.
5. Allow more time for understanding your customers' needs and fulfilling them.
6. Work at becoming a responsible and mature customer/supplier.
7. Hold a prestart 'away day' for the whole team.

Improving learning and knowledge transfer

GROUP A - Actions you can take within your own company

1. Seek to help your suppliers with learning programmes.
2. Become an advocate of good practice and new ideas.
3. Support external bodies such as CPN and the RCF with their initiatives.
4. Encourage and support effective learning within your organisation. Instill a culture of continuous improvement
5. Develop skills in problem-solving and in the use of process improvement tools.
6. Learn to work more effectively in teams.
7. Work to improve IT skills.
8. Learn to Learn
9. Learn more about your markets, clients, consultants and suppliers.
10. Seek to share your learning with your partners.
11. Undertake joint research and development
12. Participate in learning networks and alliances.
13. Learn to be an effective leader or follower.

GROUP B - Actions customers and other partners can take to make Specialist Contractors more effective

1. Encourage and help specialist contractors to develop a culture of learning and continuous improvement.
2. Encourage specialist contractors to learn to play a more proactive role in the design and construction process by raising their internal capabilities.
3. Provide opportunities for specialist contractors to develop their communication and inter-personal skills.
4. Advise all your specialist contractors and suppliers to read, digest and adopt the findings of this report.

GROUP C - Actions you can take which will need input and help from your customers and other partners

1. Share your learning and best practice with your partners.
2. Undertake joint learning programmes with your partners.
3. Undertake joint R&D projects with your partners.
4. Develop project feedback and benchmarks in order to transfer learning between projects.
5. Collaborate in the development of process maps which will become your vehicle for continuous improvement.

References

1. Latham, Sir Michael, (1994), '**Constructing the Team**' HMSO.
2. Office of Science and Technology, (1995), '**Technology Foresight: Progress Through Partnership**', HMSO.
3. Construction Industry Research and Information Association, (1997), '**Specialist Trade Contracting - A Review**', Special Publication 138
4. CBI and DTI, (1994), '**Competitiveness - How the Best UK Companies are Winning**'
5. TEC National Council, (1995), '**Manufacturing Winners - Creating a World-Class Manufacturing Base in the UK**'
6. London Business School, Severn Trent, and University of Warwick, (1995), '**Service in Britain**' .
7. Hines, P., (1994), '**Creating World-Class Suppliers**', London: Pitman.
8. Tuckman, B. W., (1965), '**Development Sequence in Small Groups**', Psychological Bulletin, Vol. 63, pp 384-399.
9. Crainer S., (1996), '**Key Management Ideas**', Pitman.
10. Bennet J ., and Jayes S ., (1995), '**Trusting the Team: The Best Practice Guide to Partnering in Construction**', Centre for Strategic Studies in Construction at the University of Reading with the Partnering Task Force of the Reading Construction Forum.
11. The Wood Report, (1975), '**The Public Client and the Construction Industries**', HMSO.
12. BAA Lynton Report, (1993), '**The UK Construction Challenge**
13. Handy C., (1995), '**Understanding Organisations**', Penguin Books.
14. Lamming, R., (1993), '**Beyond Partnership: Strategies for Innovation and Lean Supply**', New York: Prentice-Hall.
15. Economy Class in Building, Issue 01, Friday 10 January, 1997, pp 31-40
16. Womack, J.P., Jones, D.J, and Roos, D., (1990) '**The Machine that Changed the World**', Rawson Associates, New York.
17. Storey, J., (1994), '**New wave Manufacturing Strategies: Organisational and Human Resource Management Dimensions**', London; Chapman
18. Womack, J. P. and Jones, D.T., 1996), **Lean Thinking: Banish Waste and Create Wealth in your Corporation**', Simon and Schunster, New York.
19. Hines, P., and Rich, N., (1997), **The Seven Value Stream Mapping Tools**, International Journal of Operations Production Management, Vol.17, No.1.
20. Berk, J., and S., (1993), '**Total Quality Management; Implementing Continuous Improvement**', Sterling Publishing.
21. Albrecht K., (1993), '**The Only Thing That Matters**', Harper Business.
22. Juran J., (1988), '**Juran on Planning for Quality**', Free Press, New York
23. Cavaleri S. and Fearon D., (1996) '**Managing in Organisations that Learn**', Blackwell.
24. Argyris, C., (1992), '**Organisational Learning**', London: Blackwell
25. Pedler M., Burgoyne J.and Boydell T., (1997), '**The Learning Company - A Strategy for Sustainable Development**',McGraw Hill.
26. Revans, R., (1983), '**The ABC of Action Learning**', Bromley, Kent: Chartwell, Bratt.
27. Garatt B., (1990), '**Creating a Learning Organisation - A Guide to Leadership Learning and Development**', Director Books.
28. Bessant J., (1990), '**Managing Advanced Manufacturing Technology: The Challenge of the Fifth Wave**', Blackwell
29. Binney G. and Williams C. (1997), '**Leaning into the Future - Changing the Way People Change Organisations**', Nicholas Brealey Publishing, London.
30. Lambert T., (1996), '**Key Management Solutions**', Pitman Publishing.
31. Waterman R., (1994), '**The Frontiers of Excellence: Learning from Companies that Put People First**', Nicholas Brealey Publishing, London.
32. Eccles T., (1994), '**Succeeding with Change: Implementing Action Driven Strategies**', McGraw-Hill
33. Kotter, J., (1995) '**Eight Steps to Transforming your Organisation**', in Leading Change: Why Transformation Efforts Fail, Harvard Business Review,March-April